Pre-Algebra
Lessons That Speak to You

**by
Fred Truong**

Copyright © 2021 by Fred Truong
All rights reserved. No part of this book may be reproduced, scanned, or distributed in any printed or electronic form without permission.
First Edition: December 2021
Printed in the United States of America

ISBN: 9798776324758

Table of Contents

Lesson 1:   Number Sense   1
Lesson 2:   Order of Operations and Exponents   8
Lesson 3:   Manipulating Negative Integers   14
Lesson 4:   Divisibility Rules   18
Lesson 5:   Prime Numbers, LCM and GCF   23
Lesson 6:   Fractions, Decimals, Rates and Percentages   29
Lesson 7:   Ratios, Proportions and More Percentages   35
Lesson 8:   Metric Units Conversion   39
Lesson 9:   Customary Units Conversion   46
Lesson 10: Variables and Expressions   52
Lesson 11: Solving Linear Equations   56
Lesson 12: More Equations and Formulas   61
Lesson 13: Word Problems   65
Lesson 14: Solving Linear Inequalities   71
Lesson 15: Introduction to Graphs   78
Lesson 16: Graphing Linear Inequalities   85
Lesson 17: Solving Systems of Equations by Graphing   90
Lesson 18: Solving Systems of Equations by Substitution   96
Lesson 19: Laws of Exponents   102
Lesson 20: Scientific Notation   107

Answer Keys   111

Pre-Algebra

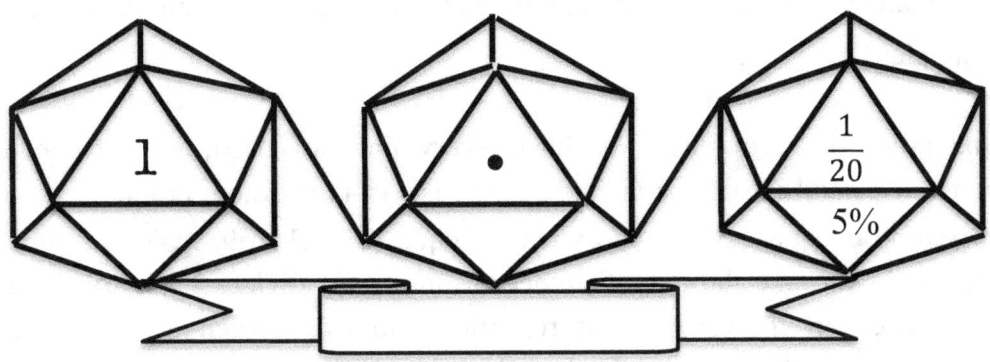

Lesson 1: Number Sense

The first step to being good at math is to develop a strong number sense. However, with the advancement in technology, students often find it difficult to resist the temptation to just enter everything into their cell phone calculators, regardless of how simple a problem maybe. Consequently, a lot of students lose not only their sense of number, but also the confidence in their ability to perform mathematical tasks at a high level.

The goal of this lesson is to help you avoid that trap and become better at doing mental math. It does not mean that from now on you can completely abandon your calculator, but rather you will rely less on it. But before I delve into computational strategies and tricks, let me first introduce the **number system**.

It is not hard to imagine that the first set of numbers humans used in their daily life were the **Natural** or **Counting numbers**. These are: $\{1, 2, 3, 4, \dots\}$

Later on, people added zero to the set of numbers, and the new system is known as the **Whole numbers**. So whole numbers are: $\{0, 1, 2, 3, 4, \dots\}$

Next, we came up with negative numbers. This new system is called the **Integer**. So integers are: $\{\dots -3, -2, -1, 0, 1, 2, 3, \dots\}$

Then, somebody thought about fractions and added them to the list. The number system is now known as the **Rational numbers**. So, rational numbers include

1

any terminating **or** repeating decimals because they can be written as fractions. For example, $1.23 = \frac{123}{100}$, or $0.3333... = \frac{1}{3}$.

At this point, it seems like we have everything we need, until someone encountered a number such as pi (the symbol is π). Pi is a non-terminating **and** non-repeating decimal. It is approximately equals to 3.14159. . . This number arrives when you take the circumference of any circle divides by its diameter. Since Pi goes on forever without repeating, we called such a number the **Irrational**. Besides pi, other examples of irrational numbers include the radicals such as $\sqrt{2}$ or $\sqrt{3}$. If you do not know what radicals are, do not worry. You will learn a lot about them in Algebra I.

Now, in math you will often hear teachers use the term **Real numbers** without explaining what they are exactly. Well, all of the above are real numbers. Are there any numbers not real? The answer is "yes," but you do not learn about **Imaginary numbers** until Algebra II. So for now, everything is real.

Ok, let's do an example.

Example 1: Classify each of the following numbers as Natural (N), Whole (W), Integer (I), Rational (R), Irrational (IR) or Real (RE).
a) 10   b) 0   c) –15   d) 5.7   e) 3π

Solution: a) It is important to notice that a number can belong in multiple sets. In fact, 10 can be classified as N, W, I, R, and RE. Also, notice that I skipped IR because a number cannot be both irrational and rational.

Here is a quick way to do these problems. You start from the top. That is, you first check to see whether a number is Natural. If it is, then you automatically write down N, W, I, R, RE. If it is not, then you move down and check to see whether it is Whole. If a number is whole, then you write W, I, R, RE. If it is not whole, you continue to move down and check.

b) 0 is not Natural, but it is Whole. So the answer is W, I, R, RE.

c) −15 is not Natural or Whole, but it is I, R, RE.

d) 5.7 is not Natural, Whole or Integer, but it is R, RE.

e) 3π is not Natural, Whole, Integer or Rational, but it is IR, RE.

Now, let's explore ways to manipulate numbers.

First, you should know how to **quickly** multiply an **integer** by 10, 100, 1000, . . . You simply attach the string of zeros to the end of the integer. For example,

$$47 \times 10 = 470 \qquad 29 \times 100 = 2900 \qquad 1000 \times 13 = 13000$$

Second, to quickly multiply or divide a **rational number** by 10, 100, 1000, . . . , you move the decimal to the right or left one or more times. For example,

$$12.59 \times 10 = 125.9 \qquad 0.431 \times 100 = 43.1 \qquad 8.57 \times 10000 = 85700$$

$$12.59 \div 10 = 1.259 \qquad 0.431 \div 100 = 0.00431 \qquad 8.57 \div 10000 = 0.000857$$

Now, these techniques are simple, right?

The problem is that students do not take advantage of them enough. You should always look for opportunities to create powers of ten in your calculations. Here are some examples. Take your time to examine them and practice.

$$\mathbf{2} \times 18 \times \mathbf{5} = 18 \times \mathbf{2} \times \mathbf{5} = 18 \times \mathbf{10} = 180$$

$$13.2 \times \mathbf{20} = 13.2 \times \mathbf{2} \times \mathbf{10} = 26.4 \times \mathbf{10} = 264$$

$$5 \times \mathbf{5} \times \mathbf{6} \times 80 = 5 \times \mathbf{6} \times \mathbf{5} \times 80 = 30 \times 400 = 12000$$

$$104 \times 22 = (100 + 4) \times 22 = \mathbf{100} \times 22 + \mathbf{4} \times 22 = 2200 + 88 = 2288$$

$$99 \times 98 = (100 - 1) \times 98 = \mathbf{100} \times 98 - \mathbf{1} \times 98 = 9800 - 98 = 9702$$

20/20 Math

Notice that in the process of making tens, sometimes I use the **commutative property of multiplication** to rearrange the order of the given numbers.

For problems in which you cannot easily make tens, especially those involving money with decimals, try to use the **estimation and compensation** technique.

Example 2: Suppose a hamburger costs $3.95 and a small fry costs $1.99, what is the total cost (excluding tax) of 3 hamburgers and 4 small fries?

Solution: Since $3.95 is close to $4, assume each hamburger is $4.
So, 3 hamburgers is $12.
Similarly, assume each small fry is $2. So, 4 small fries is $8.
The total cost is 12 + 8 = $20.
But, since we overestimated each burger by 5 cents, three burgers equal 15 cents.
Also, since we overestimated each small fry by 1 cent, four small fries equal 4 cents.
$20.00 - $0.15 - $0.04 = **$19.81**.

Notice that even at the last step, you could use estimation to quickly get the answer. $20.00 - $0.19 = **$20.00 - $0.20** + $0.01 = **$19.80** + $0.01 = $19.81.

Ok, let's finish this lesson by adding some mental math tricks to your repertoire.

I. The "Eleven" trick.

To multiply a two-digit number by 11, you can simply add the two digits and insert the sum in the middle of the two digits. For example,

$$53 \times 11 = 583 \text{ because } 5 + 3 = 8.$$

If the sum is more than 10, start from the right and "carry" over. For example,

$$86 \times 11 = 946,$$

because 8 + 6 = 14, so you put down 4, carry the 1 over to the 8 and write down 9.

Pre-Algebra

This trick works for large numbers as well. For example,

$$358162 \times 11 = 3{,}939{,}78\mathbf{2}.$$

Start from right to left. First write down a **2**, then $2 + 6 = \mathbf{8}$, $6 + 1 = \mathbf{7}$, then $1 + 8 = \mathbf{9}$, $8 + 5 = 13$ (write down **3** and carry the *1* over), $5 + 3 + \mathit{1} = \mathbf{9}$. Finally, put down the 3.

With a little practice, you will be able to quickly multiply any number by 11.

II. The "Twenty Five" trick.

*To multiply a number by 25, divide that number by **4** and attach two zeros at the end.* For example,

$$84 \times 25 = \mathbf{21}00, \text{ because } 84 \div 4 = 21.$$

What is 236×25? The answer is . . .

Well since $236 \div 4 = 59$, $236 \times 25 = 5900$. By the way, when you do $236 \div 4$, avoid setting up the division problem, do $200 \div 4$ plus $36 \div 4$ in your head.

III. The "Ten and Ten" trick.

*To multiply 2 two-digit numbers, where both numbers have the **same ten digit** and their **unit digits add to 10**, simply multiply the first digit by one plus itself and attach the product of their second digits at the end.* For examples,

$$\underline{62} \times \underline{68} = 42\mathbf{16} \text{ because } 6 \times 7 = 42 \text{ and } 2 \times 8 = 16.$$

$$\underline{93} \times \underline{97} = 90\mathbf{21} \text{ because } 9 \times 10 = 90 \text{ and } 3 \times 7 = 21.$$

If the unit digits are 1 and 9, add a "0" as a place holder. For example,

$$\underline{21} \times \underline{29} = 6\mathbf{09} \text{ because } 2 \times 3 = 6 \text{ and } 1 \times 9 = 09.$$

If you do not add the "0" place holder, it is clearly WRONG because $\underline{21} \times \underline{29}$ is NOT equal to 69. It is too small.

Now let's apply this trick to more problems.

$$15 \times 15 = 225 \qquad 25 \times 25 = 625 \qquad 35 \times 35 = 1225$$

$$45 \times 45 = 2025 \qquad 55 \times 55 = 3025 \qquad 65 \times 65 = 4225$$

$$75 \times 75 = 5625 \qquad 85 \times 85 = 7225 \qquad 95 \times 95 = 9025$$

Since any odd integer divide by 2 will end in a 5, we can do more. For examples,

$$\frac{13}{2} \times \frac{13}{2} = 6.5 \times 6.5 = 42.25$$

$$\frac{15}{2} \times \frac{15}{2} = 7.5 \times 7.5 = 56.25$$

$$\frac{19}{2} \times \frac{19}{20} = 9.5 \times 0.95 = 9.025.$$

Now, clearly it is impossible to cover everything you need to know about mental math in just one lesson. But do not worry, you will see more useful tips and tricks throughout the book. For now, you should practice what you have learned in this lesson and continue to use it in future lessons until it becomes a good habit. Have fun!

Practice 1

Classify each of each of the following numbers as Natural (N), Whole (W), Integer (I), Rational (R), Irrational (IR) or Real (RE).

1a) 100.5   1b) 16.0   1c) 0

2a) $\frac{\pi}{2}$ 2b) 1.232332333…   2c) $1.3\overline{945}$

3a) $5\frac{7}{8}$ 3b) −49.50000001   3c) -11.1π

Use mental math to compute each of the following.

4) 72 × 11   5) 32 × 25   6) 7.5 × 75

7) 123.4 × 11   8) 40 ÷ 9 × 8 × 0   9) 43 × 47 − 20

10) $\frac{17}{2} \times \frac{17}{2}$ 11) 5 × 0.06 × 80   12) 13 × 17 − 12 × 18

13) 102 × 15   14) 12.3 × 300   15) 364 ÷ 7

16) 4560 ÷ 8   17) 125 × 12   18) 105 × 105

19) At the supermarket, Mrs. Anderson bought a fish that weighs 4.2 pounds, and the unit price is $4.80 per pound. If Mrs. Anderson gives the cashier a $20 bill and a $5 bill, how much change does she get back? (Ignore tax).

20) Maria is having a pizza party in her history class this Friday. Her teacher decided to order five large pizzas and four 1-liter bottles of soda. Suppose each pizza costs $7.99 and a bottle of soda costs $2.25. How much change does Maria's teacher get back if she tips the delivery person $10, and she paid with a $100 bill? (Ignore tax).

Bonus: Compute using mental math. Then, describe the pattern.

     9 × 8   9 × 88   9 × 888   9 × 8888   9 × 88888

Lesson 2: Order of Operations and Exponents

   To further develop the number sense that you have acquired in lesson 1, you need to master the order of operations. The acronym you will need to remember is **PEMDAS** or **P**lease **E**xcuse **M**y **D**ear **A**unt **S**ally. It means Parentheses, Exponents, Multiplication/Division, Addition/Subtraction.

Notice that I put Multiplication and Division together. It is because these two operations have the same priority. That means, if you see both multiplication and division in a problem, you need to perform the operations from left to right. It is NOT always multiplication before division, as PEMDAS suggests. Similarly, addition is NOT always before subtraction.

Example: If you have $3 \times 8 \div 4$, you do $3 \times 8 = 24$ first. Then $24 \div 4 = 6$. However, if you have $32 \div 8 \times 2$, you do $32 \div 8 = 4$ first. Then $4 \times 2 = 8$. It is WRONG if you do multiplication first to get $8 \times 2 = 16$. Then $32 \div 16 = 2$.

Also, if you are not familiar with exponents, it is just a short cut for multiplication. Here are examples of how to read them and what they means.

Expression	How to read it	What it means
3^2	five **squared** or five to the second power.	$3 \times 3 = 9$
10^3	ten **cubed** or ten to the third power.	$10 \times 10 \times 10 = 1000$

2^4	two to the fourth power	$2 \times 2 \times 2 \times 2 = 16$
1^5	one to the fifth power	$1 \times 1 \times 1 \times 1 \times 1 = 1$

Note that 3^2 is NOT $3 \times 2 = 6$. That is a common mistake when you are new to exponent. The exponent 2 means you are supposed to multiply the **base** 3 by itself two times. That is why $3^2 = 9$.

In fact, here is a list of perfect squares that you will need to memorize sooner or later as you are getting into higher math.

$1^2 = 1$ $2^2 = 4$ $3^2 = 9$ $4^2 = 16$ $5^2 = 25$

$6^2 = 36$ $7^2 = 49$ $8^2 = 64$ $9^2 = 81$ $10^2 = 100$

$11^2 = 121$ $12^2 = 144$ $13^2 = 169$ $14^2 = 196$ $15^2 = 225$

Clearly, this list goes on indefinitely. So the more you know, the better. Remember: The goal is NOT to do the minimum required. If you want to get better at math, you need to spend time exploring beyond the basic. But, you should at least know perfect squares up to 15.

As you can see, the first 12 come directly from the multiplication table that you are supposed to memorize in elementary school.

For 13^2 and 14^2, it is easy to remember because they are similar (you just need to switch the ten and the unit digits).

As for 15^2, you already learned the trick in lesson 1. Here they are again (in exponent form). I also added 105^2 and 115^2 to the list.

$15^2 = 225$ $25^2 = 625$ $35^2 = 1225$ $45^2 = 2025$

$55^2 = 3025$ $65^2 = 4225$ $75^2 = 5625$ $85^2 = 7225$

$95^2 = 9025$ $105^2 = 11025$ $115^2 = 13225$

Now let's do some examples with order of operations.

20/20 Math

Example 1: Simplify each expression.

a) $18 + 27 \div 9$ b) $16 - 8 \times 3 \div 4 + 11$ c) $2 \times 6^2 \div 8$

d) $(6 + 7)^2 - 40 \times 3$ e) $200 - [2 \times (18 - 11)]^2$

Solution: a) Warning: If you enter this expression one by one into your calculator, you may or may not get the right answer. It is depends on the type of calculator you use. So be careful.

According to PEMDAS, you need to do division before addition. So $27 \div 9 = 3$. Then, $18 + 3 = \mathbf{21}$.
If you do it from left to right, you will get 5, which is wrong.

b) First, do $8 \times 3 = 24$. Then, do $24 \div 4 = 6$.
Now you have $16 - 6 + 11$. Since addition and subtraction have the same priority, you do it from left to right.
$16 - 6 = 10$. $10 + 11 = \mathbf{21}$.

c) First, do $6^2 = 36$. Then, do $2 \times 36 = 72$. Finally, $72 \div 8 = \mathbf{9}$.
It is a common mistake to do $2 \times 6 = 12$ first, and then do 12^2. Since there is no parenthesis, the exponent only belong to the 6, not the 2.

d) Here you have to do the parentheses first. So $(6 + 7) = 13$. Then $13^2 = 169$. Next, multiplication goes before subtraction; so $40 \times 3 = 120$. Finally, $169 - 120 = \mathbf{49}$.

e) Notice here we have a pair of brackets. When there are too many parentheses, we use brackets to make it looks clear and different from the parentheses. You can think of them as just another pair of parentheses.

So first you have to do the inner parentheses, $(18 - 11) = 7$.
Then you do the outer parentheses or brackets, $[2 \times 7] = 14$.
Next is the exponent, $14^2 = 196$.
Finally, $200 - 196 = \mathbf{4}$.

Example 2: Compare. Write <, >, or =.

    a) $38 - 27 + 9$ __ $38 - (27 + 9)$

    b) $(2 + 3^3) \times 10$ __ $(2 + 3)^3 \times 10$

    c) $25 \times 2^5 \div 8$ __ $25 \times (2^5 \div 8)$

Solution: a) If you have $38 - 27 + 9$, you have to do it from left to right. So the answer is $11 + 9 =$ **20**. As for $38 - (27 + 9)$, you have to do the parentheses first. So the answer is $38 - 36 =$ **2**, and **20 > 2**.

b) If you have $(2 + 3^3) \times 10$, you have to go inside the parentheses first, and then do $3^3 = 27$. After that, $(2 + 27) = 29$. $29 \times 10 =$ **290**. As for $(2 + 3)^3 \times 10$, you also have to go inside the parentheses first, and then do $(2 + 3) = 5$. After that, $5^3 = 125$. Finally, $125 \times 10 =$ **1250**, and **290 < 1250**.

c) With $25 \times 2^5 \div 8$, you have to do $2^5 = 2 \times 2 \times 2 \times 2 \times 2 = 32$. Then, $25 \times 32 = 800$ (remember the "25 trick"?).
Finally, $800 \div 8 =$ **100**.
As for $25 \times (2^5 \div 8)$, you have to go inside the parentheses, and do $(2^5 \div 8) = (32 \div 8) = 4$. Finally, $25 \times 4 =$ **100**, and **100 = 100**.

Example 3: Insert grouping symbols to make each statement true.

    a) $6 \times 2 + 9 \div 3 = 22$

    b) $7 + 5^2 \div 4 = 36$

    c) $7 + 8 \times 3^4 - 79 = 30$

Solution: a) Grouping symbols means parentheses. For these problems, they are more challenging, and so you may have to try to insert parentheses into several places before you get the right answers. For $6 \times 2 + 9 \div 3$, insert the parentheses around $2 + 9$ will gives you $6 \times (2 + 9) \div 3 = 6 \times 11 \div 3 = 66 \div 3 = 22$.

b) This problem is especially difficult. So if you get it, you did a good job. In order to get 36, you need to insert the parentheses around $7 + 5$. This way, $(7 + 5)^2 = 12^2 = 144$, and $144 \div 4 = 36$.

c) To get 30 here, you need to insert parentheses around $7 + 8$ and also $3^4 - 79$. This way, we have $(7 + 8) \times (3^4 - 79) = 15 \times 2 = 30$.

Note: To calculate 3^4 in your head, DO NOT do $3 \times 3 = 9$, $9 \times 3 = 27$, and $27 \times 3 = 81$. It is faster to multiply them in pairs as follows: $3^4 = 3 \times 3 \times 3 \times 3 = 9 \times 9 = 81$. Here is another example, $2^7 = (2 \times 2) \times (2 \times 2) \times (2 \times 2) \times 2 = 4 \times 4 \times 4 \times 2 = 64 \times 2 = 128$.

Ok that's it! It's time for you to practice again. However, if you still feel a little unclear about any parts of this lesson and/or the previous lesson, please re-read them at least one more time. You need to understand and remember everything before you move on. Also, reading math is different from reading an English novel. Take your time and read slowly. Let important mathematical ideas sink in. In other words, **keep wonder and ponder**!

Practice 2

Simplify each expression.

1) $207 - 17 \times 11$

2) $30 + 25 \times 8 - 1000 \div 20$

3) $(11 - 7)^3 - 8 \times 3$

4) $2 \times 25^2 \div 100$

5) $12 \times 18 \div 6^3$

6) $1500 \div [(23 - 16) \times 5]^2$

7) $3 \times \{5 - 3 \times [7 - 3 \times (5 - 3)] + 1\}$

8) $[\frac{1}{4} \times (98 \div 49)^4]^3$

Compare. Write <, >, =.

9) $280 \div 7 \times 2$ ___ $280 \div (7 \times 2)$

10) $32 + (4 \times 5^2)$ ___ $32 + (4 \times 5)^2$

11) $(15 + 4) \times (6 - 3)$ ___ $15 + 4 \times 6 - 3$

12) $120 + 2^7 - 8$ ___ $120 + (2^7 - 8)$

13) $101 - 49 - 2.5$ ___ $101 - (49 - 2.5)$

14) $46 \div (2 \div 2)$ ___ $46 \div 2 \div 2$

Insert grouping symbols to make each statement true.

15) $40 \div 8 - 3 \times 12 = 96$

16) $2 + 3 + 4 \times 5 - 6 = 31$

17) $88 - 8 \div 8^2 - 48 = 5$

18) $16 + 8 \div 6 \times 4 = 16$

19) $206 - 9 - 3^2 = 170$

20) $3 + 3 + 3 \times 3 - 3 \div 3 = 20$

Bonus: Compute each of the following using mental math.

205^2 305^2 405^2 505^2 605^2 705^2 805^2 905^2

20/20 Math

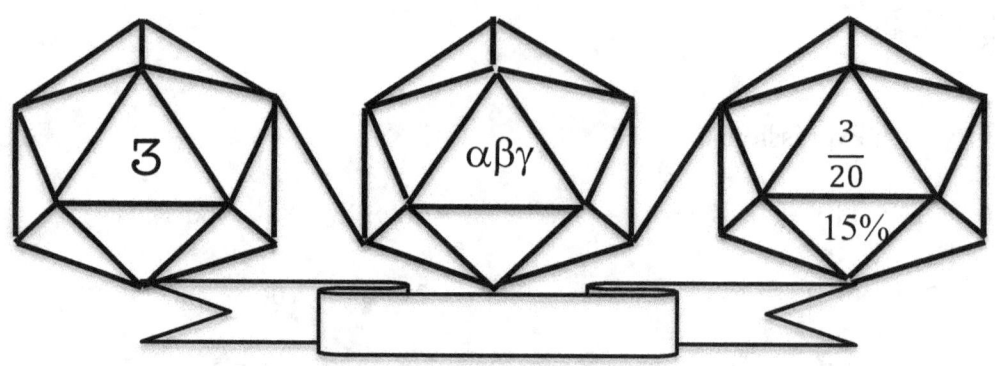

Lesson 3: Manipulating Negative Integers

Up to this point, you have only worked with positive numbers. This lesson will help you to expand your knowledge of negative numbers.

Important: To add or subtract, think of negative integers as you "**owe**" someone money and positive integers as you "**have**" money.

Example 1: Add or Subtract.

        a) $-7 + 5$ b) $-3 + 9$ c) $-2 - 4$

        d) $5 - 9$ e) $10 - 13 - 6$

Solution: a) $-7 + 5$ means you **owe 7**, and you **have 5**. After you paid the 5, you still owe 2. So, the answer is **–2**.

b) $-3 + 9$ means you **owe 3**, and you **have 9**. The result is **6**.

c) $-2 - 4$ means you **owe 2**, and you **owe 4 more**. So, you owe a total of 6. The answer is **–6**.

d) $5 - 9$ means you **have 5**, and you **owe 9**. So, you still owe 4. The answer is **–4**.

e) 10 – 13 – 6 means you **have 10**; you **owe 13**. So, you still **owe 3**. Then, you **owe 6 more**. So, you owe 9 total. The answer is **–9**.

Important: When you see "+ (–)" or "– (+)", ignore the "+" sign.
When you see "– (–)", turn them into the "+" sign.

Example 2: Add or Subtract.

    a) –16 + (–9)   b) 12 + (–23)   c) 13 – (–4)

    d) –7 – (–19)   e) 14 – (–8) + (–36)

Solution: a) –16 + (–9) = –16 – 9, because you would ignore the "+" sign. This means you **owe 16**, and you **owe 9 more**. So, you owe 25 total. The answer is **–25**.

b) 12 + (–23) = 12 – 23. This means you **have 12**, and you **owe 23**. So, you owe 11 at the end. The answer is **–11.**

c) 13 – (–4) = 13 + 4 = 17, because "– (–)" means "+".

d) –7 – (–19) = –7 + 19. This means you **owe 7**, and you **have 19**. So, you have 12 at the end. The answer is **12.**

e) 14 – (–8) + (–36) = 14 + 8 – 36 = 22 – 36. This means you **have 22**, and you **owe 36**. So, you owe 14. The answer is **–14**.

Important: To multiply or divide integers, use the following simple rules:
If two numbers have the **same sign**, the product or quotient is **positive**.
If two numbers have **different signs**, the product or quotient is **negative**.

Example 3: Multiply or Divide.

    a) –8 × (–9)   b) 21 × (–20)   c) –132 ÷ (11)

    d) $(-15)^2$ e) $-42 \div 7 \times (-2)^3$

Solution: a) –8 × (–9) = 72, because negative × negative = positive.

b) $21 \times (-20) = -420$, because positive × negative = negative.

c) $-132 \div (11) = -12$, because negative × positive = negative.

d) $(-15)^2 = (-15) \times (-15) = 225$, because neg. × neg. = pos.

e) First, $(-2)^3 = (-2) \times (-2) \times (-2) = 4 \times (-2) = -8$.
So, $-42 \div 7 \times (-2)^3 = -42 \div 7 \times (-8) = -6 \times (-8) = 48$.

Example 4: Compare. Write <, >, or =.

a) $-7 + 19$ __ -26

b) $14 + (-14)$ __ 0

c) $75 - (-75)$ __ 0

d) $(-2)^5$ __ $(-2)^2$

e) $10 - 20 - 30$ __ $10 - (20 - 30)$

Solution: a) $-7 + 19$ means you **owe 7**, and you **have 19**. So, you still have **12**, and **12 > −26**.

b) $14 + (-14) = 14 - 14 = \mathbf{0}$.

c) $75 - (-75) = 75 + 75 = 150$. So, **150 > 0**.

d) Even though 5 is more than 2, it is an odd number. A negative number raises to an odd number exponent is negative. In fact, $(-2)^5 = (-2)(-2)(-2)(-2)(-2) = (4)(4)(-2) = 16 \times (-2) = -32$. Whereas, $(-2)^2 = (-2)(-2) = 4$, and **−32 < 4**.

e) For $10 - 20 - 30$, you have to do it from left to right. $10 - 20 = -10$. Then, $-10 - 30 = \mathbf{-40}$.
For $10 - (20 - 30)$, you have to do inside the parentheses first. $20 - 30 = -10$. Then, $10 - (-10) = 10 + 10 = \mathbf{20}$.
So, **−40 < 20**.

Practice 3

Add or Subtract.

1) $-18 + 36$

2) $21 - 40$

3) $-1000 - 1000$

4) $-10 - (-100)$

5) $-50 + (-100)$

6) $11 - 21 - 31$

Multiply or Divide.

7) -80×25

8) $-196 \div -14$

9) $-34 \div (-2) \times (-11)$

10) $(-9)^3 \div (-10)$

11) $5 \times (-87) \times (-2)$

12) $(-5)^4 \div (-10)^2$

Simplify.

13) $30 \div (4 - 9) \times (-36)$

14) $[-2 + (3 - 4) \times 5]^2 - 6$

15) $[-88 - (-8)] + (48 - 8^2)$

16) $[(-12 - 18) \div 6] \times 20$

Compare. Write <, >, =.

17) -23×18 __ $23 \times (-18)$

18) $[-14 - (-14)]^{14} - 14 \div 14$ __ 0

19) $100 - (7 - 77)$ __ $100 - 7 - 77$

20) $-92 \div (2 - 4)^2$ __ -22

Bonus: Simplify.

$$(-1)^2 + (-1)^3 + (-1)^4 + (-1)^5 + (-1)^6 + (-1)^7 + (-1)^8 + (-1)^9$$

Lesson 4: Divisibility Rules

Part of having a good number sense is knowing whether one number is divisible by another number. Fortunately, there are rules that allow you to quickly determine if a number is divisible by 2, 3, 4, 5, 6, 8, 9, or 10.

	A number is divisible by
2	if it is an **even** number (i.e., it ends in 0, 2, 4, 6, or 8). **For example:** 1354 is divisible by 2 since it ends in a "4".
3	if the **sum of the digits** is divisible by 3. **For example:** 1725 is divisible by 3 because 1 + 7 + 2 + 5 = 15, and 15 is divisible by 3.
4	if the **last two digits** is divisible by 4. **For example:** 3516 is divisible by 4 because 16 is divisible by 4.
5	if it ends in a "**0**" or a "**5**". **For example:** 935 is divisible by 5 because it ends in a "5".
6	if it divisible by both **2 and 3**. **For example:** 348 is even, so it is divisible by 2. 3 + 4 + 8 = 15, so it is divisible by 3. Therefore, 348 is divisible by 6.

8	if **the last three digits** is divisible by 8. **For example:** 235160 is divisible by 8 because 160 is divisible by 8.
9	if **the sum of the digits** is divisible by 9. **For example:** 5193 is divisible by 9 because $5 + 1 + 9 + 3 = 18$, and 18 is divisible by 9.
10	if it is ends a "**0**". **For example:** 27890 is divisible by 10 because it ends in a "0".

Example 1: Determine whether the first number is divisible by the second.

    a) 5015; 3   b) 4698; 9   c) 4150; 4

    d) 642; 8   e) 5342; 6

Solution: a) $5 + 0 + 1 + 5 = 11$, and 11 is NOT divisible by 3. Hence, 5015 **is NOT** divisible by 3.

b) $4 + 6 + 9 + 8 = 27$, and 27 is divisible by 9. Hence, 4698 **is** divisible by 9.

c) The last two digits of 4150 is 50, and 50 is NOT divisible by 4. Hence, 4150 **is NOT** divisible by 4.

d) The rule for 8 is, you have to look at the last three digits and see if it divisible by 8. This rule is only useful if the number has more than three digits. With 642, you actually have to divide by 8 to see if it is divisible by 8. However, to practice using your number sense, you can reason as follow: $64 \div 8 = 8$. $640 \div 8 = 80$. Hence, $642 \div 8 = 80$ Remainder 2. It **is NOT** divisible by 8.

e) 5342 is divisible by 2 because it is an even number. $5 + 3 + 4 + 2 = 14$, and 14 is NOT divisible by 3. That means 5342 is NOT divisible by 3. Hence, 5342 **is NOT** divisible by 6.

Example 2:   Determine if each number is divisible by 2, 3, 4, 5, 6, 8, 9, or 10.

   a) 543210   b) 123123   c) 222444   d) 97

Solution: a) Since 543210 ends in a "0", it is divisible by **2, 5, and 10**.
$5 + 4 + 3 + 2 + 1 + 0 = 15$. So, 543210 is divisible by **3**, but it is NOT divisible by **9**. Also, 543210 is divisible by **6** because it is divisible by both 2 and 3. 543210 is NOT divisible by **4** because 10 is NOT divisible by 4. 543210 is NOT divisible by **8** because 210 is NOT divisible by 8.

   b) Since 123123 ends in a "3", it is NOT divisible by **2, 5, or 10**. Since 321321 is NOT divisible by 2, it is also NOT divisible by **6**. $1 + 2 + 3 + 1 + 2 + 3 = 12$. So, 123123 is divisible by **3**, but it is NOT divisible by **9**. 123123 is NOT divisible by **4** because 23 is NOT divisible by 4. 123123 is NOT divisible by **8** because 123 is NOT divisible by 8.

   c) Since 222444 ends in a "4", it is divisible by **2,** but NOT divisible by **5, or 10**.
$2 + 2 + 2 + 4 + 4 + 4 = 18$. So, 222444 is divisible by **3** and **9**. 222444 is divisible by **6** because it is divisible by both 2 and 3. 222444 is divisible by **4** because 44 is divisible by 4.
Finally, 222444 is NOT divisible by **8** because 444 is NOT divisible by 8. (Note that $440 \div 8 = 55$).

   d) Since 97 ends in a "7", it is NOT divisible by **2, 5, or 10**. Since 97 is NOT divisible by 2, it is also NOT divisible by **6**. $9 + 7 = 16$. So, 97 is NOT divisible by **3** or **9**. 97 is NOT divisible by **4** because $25 \times 4 = 100$ and $24 \times 4 = 96$. 97 is NOT divisible by **8** because $12 \times 8 = 96$.
In fact, 97 is NOT divisible by anything but 1 and itself. It is a **prime** number.

Example 3:   Find a number that satisfies the given conditions.

   a) A five-digit number divisible by 2 and 5, but not divisible by 9.
   b) A number divisible by 2, 3, 4, and 6, but not divisible by 8.

Pre-Algebra

c) A number divisible by 2, 3, 4, 5, 6, 8, 9, and 10.

Solution: a) To be divisible by 5, a number must end in a "0" or "5". But to be divisible by 2, a number **cannot** end in a "5". So, the five-digit number you are looking for **must end** in a "0". Also, the sum of its digits cannot be divisible by 9. One possible answer is **12340**. A second possible answer is **43210**.

b) To be divisible by 2 and 4, the last two digits must be divisible by 4. For example, the last two digits could be 12. To be divisible by 3, the sum of the digits must be divisible by 3. 12 still works in this case. If the number is divisible by 2 and 3, it automatically divisible by 6. Finally, 12 is NOT divisible by 8. Hence, **12** is a possible answer.

c) A number ends in "0" is divisible by **2, 5 and 10**. If the sum of the digits of the number is divisible by 9, then the number is divisible by **3 and 9**. If the number is divisible by both 2 and 3, then it divisible by **6**. Finally, if the last three digits of the number is divisible by 8, then the number is divisible by **4 and 8**. One possible answer is **2160**. A second possible answer is **92160**. A third possible answer is **992160**. The reasoning here is, by adding an extra "9" into the front of the number, the sum of the digits stay divisible by 9, and it does not affect the last three digits.

Practice 4

Determine whether the first number is divisible by the second number.

1) 111222333; 2

2) 75312; 3

3) 9834; 4

4) 10209; 5

5) 135246; 6

6) 32084; 8

7) 3,456,789; 9

8) 1,010,101; 10

9) 9008; 8

10) 10736; 6

Determine if each number is divisible by 2, 3, 4, 5, 6, 8, 9, or 10.

11) 2020

12) 91

13) 777555

14) 964

15) 1400

16) 10000

Find a number that satisfies the given conditions.

17) A four-digit number divisible by 2, 4, 6 and 10, but not divisible by 8.

18) A five-digit number divisible by 3, 5 and 8, but not divisible by 9.

19) A number divisible by 3 and 5, but not divisible by 6.

20) A three-digit number divisible by 4 and 6, but not divisible by 8.

Bonus: Without multiplying, determine if 12345 × 6789 is divisible by 2 or 5.

Pre-Algebra

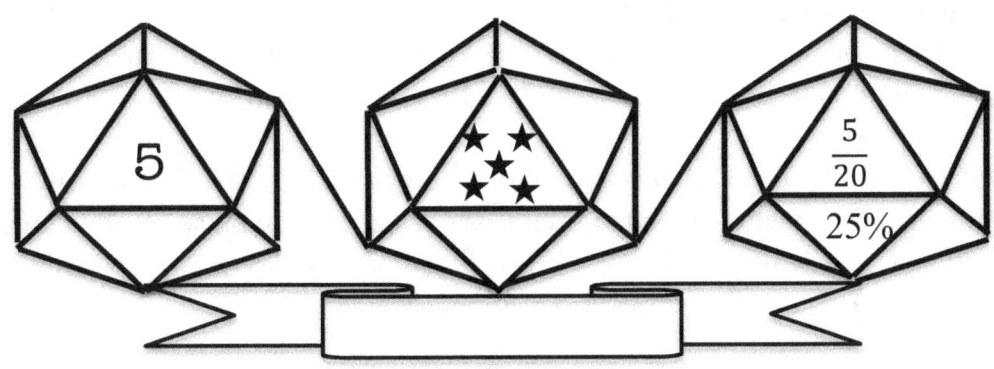

Lesson 5: Prime Numbers, LCM and GCF

Recall that a **prime** number is a number that is only divisible by one and itself. If a number is not prime, then it is a **composite** number, except "one" is neither prime nor composite. Two is the only even prime number. In fact, all prime numbers up to 100 are underlined in the table below. You should know them.

1	**2**	**3**	4	**5**	6	**7**	8	9	10
11	12	**13**	14	15	16	**17**	18	**19**	20
21	22	**23**	24	25	26	27	28	**29**	30
31	32	33	34	35	36	**37**	38	39	40
41	42	**43**	44	45	46	**47**	48	49	50
51	52	**53**	54	55	56	57	58	**59**	60
61	62	63	64	65	66	**67**	68	69	70
71	72	**73**	74	75	76	77	78	**79**	80
81	82	**83**	84	85	86	87	88	**89**	90
91	92	93	94	95	96	**97**	98	99	100

Now, the process of breaking down a composite number into factors of prime numbers is called **prime factorization**.

For example, to find the prime factorization of 60, start by breaking 60 down into 6 × 10. Then continue to break 6 down into 2 × 3 and 10 into 2 × 5. The prime factorization of 60 is **2 × 3 × 2 × 5 = 2 × 2 × 3 × 5 = 2² × 3 × 5**.

What if you start by breaking 60 down into 2 × 30 instead of 6 × 10? You will get the same prime factorization at the end, as you continue to break 30 down into 5 × 6. Finally, break 6 down into 2 × 3. So, how you would start does not affect your final result.

Here are the two "factor trees" for the above prime factorization of 60.

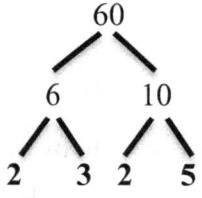
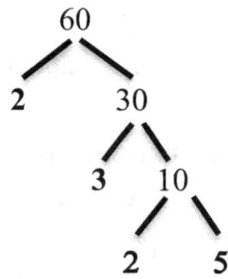

Example 1: Write the prime factorization of each number.

        a) 124   b) 192   c) 360

Solution: a) 124 = 2 × 62 = **2 × 2 × 31**. Note that 31 is a prime number.

b) 192 = 3 × 64 = 3 × 2 × 32 = 3 × 2 × 2 × 16 = 3 × 2 × 2 × 2 × 8 = 3 × 2 × 2 × 2 × 2 × 4 = **3 × 2 × 2 × 2 × 2 × 2 × 2 = 3 × 2⁶**.

c) 360 = 36 × 10 = 6 × 6 × 2 × 5 = **2 × 3 × 2 × 3 × 2 × 5** = **2³ × 3² × 5**.

Using prime factorizations, you can find the Least Common Multiple (LCM) and the Greatest Common Factor (GCF) of two or more numbers.

Example 2: Find the Least Common Multiple (LCM).

        a) 36, 120   b) 18, 24   c) 12, 15, 20

Solution: a) It is not uncommon that some students think the answer is 12 for this problem. If you are one of those students, you mistakenly find the GCF instead of the LCM. The confusion stems from the words "Least" and "Greatest". To some students, they associate the "Least" Common Multiple with a "smaller" number, and the "Greatest" Common Factor with a "bigger" number. That is incorrect. So, how do we fix it?

First and foremost, remember my new acronym from now on.

LCM = Let's Collect More!!!

That means to find the LCM of 36 and 120, you begin by doing the prime factorization of 36 and 120 as usual. So, here they are:
$36 = 6 \times 6 = 2 \times 3 \times 2 \times 3 = \mathbf{2^2 \times 3^2}$.
$120 = 12 \times 10 = 2 \times 6 \times 2 \times 5 = 2 \times 2 \times 3 \times 2 \times 5 = \mathbf{2^3 \times 3 \times 5}$.

Now "Let's Collect More." Since there are **two** factors of 2 in 36 and **three** factors of 2 in 120, let's **collect 2 × 2 × 2**.
Similarly, there are **two** "**3**" in 36, but only **one** "**3**" in 120, let's **collect 3 × 3**. Finally, there is **zero** "**5**" in 36, but **one** "**5**" in 120, let's **collect 5**. So, the LCM is **2 × 2 × 2 × 3 × 3 × 5 = 360**.

b) $18 = 3 \times 6 = 3 \times 2 \times 3 = \mathbf{2 \times 3^2}$.
$24 = 2 \times 12 = 2 \times 2 \times 6 = 2 \times 2 \times 2 \times 3 = \mathbf{2^3 \times 3}$.

Notice there are **one** "**2**" and **two** "**3**" in 18, and there are **three** "**2**" and **one** "**3**" in 24. Let's collect **2 × 2 × 2 × 3 × 3**. So, the LCM is **72**.

c) $12 = 6 \times 2 = 3 \times 2 \times 2 = \mathbf{2^2 \times 3}$.
$15 = \mathbf{3 \times 5}$.
$20 = 10 \times 2 = 5 \times 2 \times 2 = \mathbf{2^2 \times 5}$.

There are **two** "**2**" in 12, **zero** "**2**" in 15, and **two** "**2**" in 20. Notice there is a tie here. The most "**2**" is two "**2**". Let's collect **2 × 2**. There is **one** "**3**" in 12, **one** "**3**" in 15, and **zero** "**3**" in 20. Let's

20/20 Math

collect one **3**.
There is **zero** "5" in 12, **one** "5" in 15, and **one** "5" in 20. Let's collect one **5**. So the LCM is **2 × 2 × 3 × 5 = 60.**

Example 3: Find the Greatest Common Factor (GCF).

  a) 32, 48   b) 56, 91   c) 49, 79   d) 18, 30, 42

Solution: a) Now that you learn LCM = Let's Collect More, remember GCF is the opposite of LCM.

GCF = Go Collect Fewer!!!

That means to find the GCF of 32 and 48, you begin by doing the prime factorization of 32 and 48 as usual. So, here they are:
32 = 4 × 8 = 2 × 2 × 2 × 4 = 2 × 2 × 2 × 2 × 2 = **2^5**.
48 = 6 × 8 = 2 × 3 × 2 × 4 = 2 × 3 × 2 × 2 × 2 = **2^4 × 3**.

There are **five** "2" in 32, but only **four** "2" in 48.
Go collect **2 × 2 × 2 × 2**.
There is **zero** "3" in 32 and **one** "3" in 48. Go collect **zero** "3".
So, the GCF is **2 × 2 × 2 × 2 = 16.**

b) 56 = 7 × 8 = 7 × 2 × 4 = 7 × 2 × 2 × 2 = **7 × 2^3**.
Now if you look at the above table, 91 is NOT a prime number. So, if you do not already know factors of 91, you have to try to find them. Based on the divisibility rules, 91 is NOT divisible by 2, 3, 4, 5, and 6. So naturally, you would have to try 7 first. As it turns out, 91 = **7 × 13**.

There is **one** "7" in 56 and **one** "7" in 91. Go collect **7**.
There is **three** "2" in 56 and **zero** "2" in 91. Go collect **zero 2**.
There is **zero** "13" in 56 and **one** "13" in 91. Go collect **zero 13**.
So, the GCF is **7**.

c) 49 = **7 × 7**. 79 = **1 × 79** because it is a prime number. (If you are not sure, you can look it up in the above table).

There are **two** "7" in 49, but **zero** "7" in 79. Go collect **zero** 7.
There are **zero** "79" in 49, and **one** "79" in 79. Go collect **zero** 79.
So, what is the GCF? Well, the answer is **1**, because 49 could be written as **1** × **7** × **7** and 79 = **1** × **79**.

d) 18 = 3 × 6 = 3 × 2 × 3 = **2 × 3²**.
30 = 3 × 10 = **3 × 2 × 5**.
42 = 6 × 7 = **2 × 3 × 7**.

There is **one** "2" in 18, **one** "2" in 30, and **one** "2" in 42. Go collect **one 2**.
There are **two** "3" in 18, **one** "3" in 30, and **one** "3" in 42. Go collect **one 3**.
There is **zero** "5" in 18, **one** "5" in 30, and **zero** "5" in 42. Go collect **zero 5**.
There is **zero** "7" in 18, **zero** "7" in 30, and **one** "7" in 42. Go collect **zero 7**. So, the GCF is **2 × 3 = 6**.

Example 4: Jenny takes her dog to the park every 8 days and cleans her room every two weeks. If she does both today, how many days will pass before she does them both on the same day again?

Solution: Since Jenny took her dog to the park today, she would do it again in 8 days, 16 days, 24 days, and etc . . . In other words, you are looking for multiples of 8.
Similarly, since Jenny cleans her room today, she would do it again in two weeks or 14 days, 28 days, 42 days, and etc . . .
To find the first time these multiples meet is the same as finding the Least Common Multiple of 8 and 14.
8 = 2 × 2 × 2 = **2³**. 14 = **2 × 7**. So, let's collect **2 × 2 × 2 × 7 = 56**. Jenny will do both activities again **56 days** from today.

20/20 Math

Practice 5

Write the prime factorization of each number.

1) 252

2) 960

3) 1350

4) 1386

Find the Least Common Multiple (LCM).

5) 28, 42

6) 8, 14

7) 160, 240

8) 97, 291

9) 40, 50, 60

10) 32, 64, 80

11) 15, 25, 30

12) 21, 26, 91

Find the Greatest Common Factor (GCF).

13) 12, 37

14) 28, 98

15) 84, 102

16) 65, 169

17) 8, 12, 27

18) 108, 117, 171

19) 96, 112, 256

20) 54, 81, 243

Bonus: Determine if 323 is prime or composite. If it is composite, write the prime factorization.

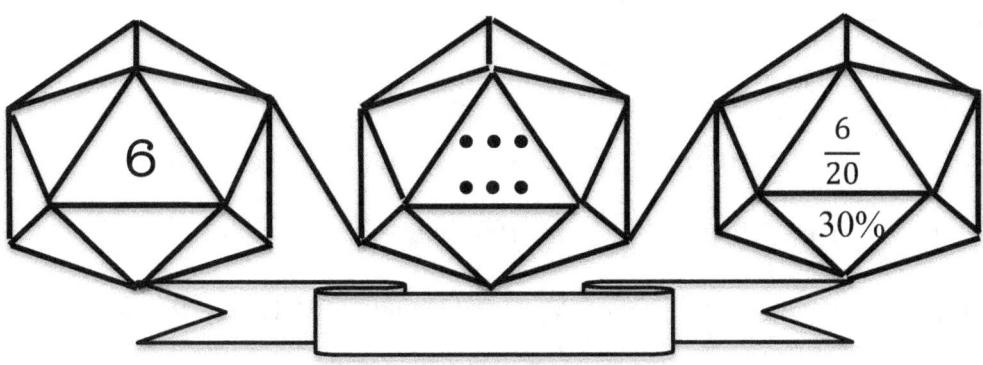

Lesson 6: Fractions, Decimals, Rates and Percentages

Fear of fractions is one of the reasons many students feel they cannot conquer math. So, the goal of this lesson is to help you become a little bit more comfortable working with fractions. Now, since this is an Pre-Algebra book, I will not take you all the way back to fourth grade and reteach you all the basics of fractions. Instead, let's start by reviewing the following facts.

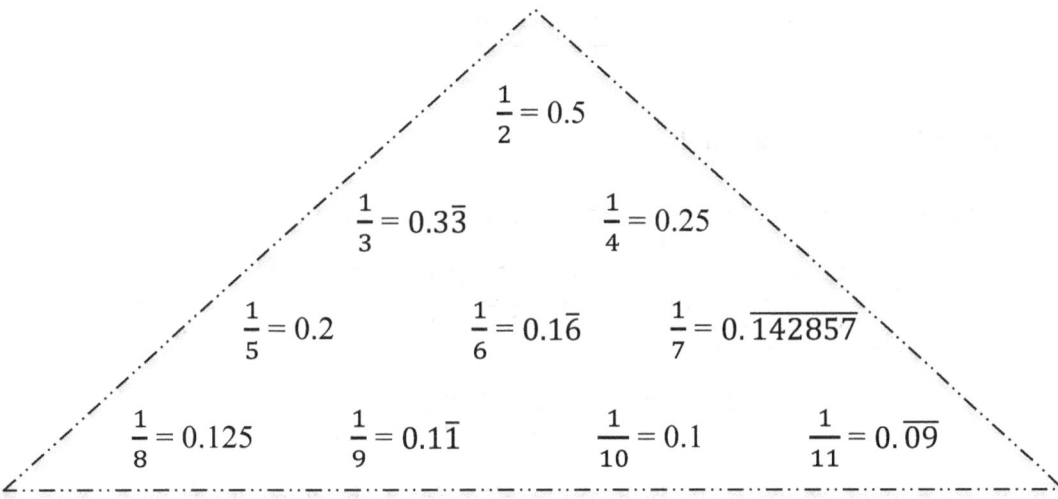

Notice that 9 out of the 10 fractions above are easy to remember (except $\frac{1}{7}$). So, you should know them. One particularly useful fact is, $\frac{1}{9} = 0.1\bar{1}$. Since $\frac{2}{9} = 2 \times \frac{1}{9}$, that means $\frac{2}{9} = 0.2\overline{2}$. Similarly, $\frac{3}{9} = 0.3\overline{3}$. In fact, you can now divide a lot of numbers by 9 quickly.

Example 1: Use mental math to find each quotient.

a) $68 \div 9$ b) $1007 \div 9$ c) $135 \div 11$

Solution: a) Since $63 \div 9 = 7$, $68 \div 9 = 7$ remainder 5 (out of 9). So,

$$68 \div 9 = 7\frac{5}{9} = 7.5\overline{5} = 7.5555\ldots$$

b) Since $999 \div 9 = 111$, $1007 \div 9 = 111$ remainder 8. So,

$$1007 \div 9 = 111\frac{8}{9} = 111.8\overline{8}$$

c) Since $132 \div 11 = 12$, $135 \div 11 = 12\frac{3}{11} = 12 + 3 \times \frac{1}{11}$

$$= 12 + 3 \times 0.\overline{09} = 12.\overline{27}$$

Given a fraction, you can multiply or divide the numerator and denominator by the same number to create **equivalent fractions**. For example,

$$\boxed{\frac{2}{3} = \frac{2}{3} \times \frac{2}{2} = \frac{4}{6}}, \text{ or } \boxed{\frac{2}{3} = \frac{2}{3} \times \frac{3}{3} = \frac{6}{9}}, \text{ or } \boxed{\frac{2}{3} = \frac{2}{3} \times \frac{x}{x} = \frac{2x}{3x}}$$

$$\frac{20}{48} = \frac{20}{48} \div \frac{2}{2} = \frac{10}{24}, \text{ or } \frac{20}{48} \div \frac{4}{4} = \frac{5}{12}$$

Using equivalent fractions, you can find **unit rates** and **percentages**.

Example 2: Find the unit rate.

a) $\dfrac{420 \text{ miles}}{5 \text{ hours}}$ b) $\dfrac{\$90}{8 \text{ people}}$ c) $\dfrac{72 \text{ grams}}{108 \text{ liters}}$

Solution: a) To find the unit rate, make the denominator equal one.

$$\frac{420 \text{ miles}}{5 \text{ hours}} = \frac{420 \text{ miles}}{5 \text{ hours}} \div \frac{5}{5} = \frac{84 \text{ miles}}{1 \text{ hour}} = 84 \text{ miles per hour}$$

b) $\frac{\$90}{8 \text{ people}} = \frac{\$90}{8 \text{ people}} \div \frac{8}{8}$

Since $88 \div 8 = 11$, $90 \div 8 = 11$ remainder 2 out of 8.

So, $\qquad 90 \div 8 = 11\frac{2}{8} = 11\frac{1}{4} = 11.25.$

Hence, $\frac{\$90}{8 \text{ people}} = \frac{\$90}{8 \text{ people}} \div \frac{8}{8} = \frac{\$11.25}{1 \text{ person}} = 11.25$ dollars per person

c) $\frac{72 \text{ grams}}{108 \text{ liters}} = \frac{72 \text{ grams}}{108 \text{ liters}} \div \frac{36}{36} = \frac{2 \text{ grams}}{3 \text{ liters}} \approx 0.67$ grams per liter

Important: While **unit** rate means "per one," **percent** means "per one hundred."

Example 3: Calculate the percent.

a) $\frac{19}{20}$ b) $\frac{21}{25}$ c) $\frac{52}{400}$ d) $\frac{42}{54}$

Solution: a) $\frac{19}{20} = \frac{19}{20} \times \frac{5}{5} = \frac{95}{100} = 95\%$

b) $\frac{21}{25} = \frac{21}{25} \times \frac{4}{4} = \frac{84}{100} = 84\%$

c) $\frac{52}{400} = \frac{52}{400} \div \frac{4}{4} = \frac{13}{100} = 13\%$

d) This time we cannot easily turn 54 into 100. So, we reduce the fraction as follow: $\frac{42}{54} = \frac{42}{54} \div \frac{6}{6} = \frac{7}{9} = 0.\overline{7777} = \frac{77.\overline{77}}{100} \approx 77.78\%$

Example 4: Write each percent as a decimal.

a) 39%   b) 237%   c) 0.38%   d) -1.6%

Solution: a) $39\% = \frac{39}{100} = 0.39$

b) $237\% = \frac{237}{100} = 2.37$

c) $0.38\% = \frac{0.38}{100} = 0.0038$

d) $-1.6\% = \frac{-1.6}{100} = -0.016$

Notice that to convert percent to decimal, you can simply move the decimal two places to the left.

Example 5:   Find each of the following.

a) 40% of 90   b) 65% of 65   c) 32% of 25   d) 900% of 8.1

Solution: a) 40% means $\frac{40}{100}$, and "of" means multiply.

So, 40% of 90 means $\frac{40}{100} \times 90 = 36$.

b) Now you should be able to do this problem in your head.

$\frac{65}{100} \times 65 = \frac{4225}{100} = 42.25$.

c) If you rearrange 32 and 25, then you can easily do this problem.

$\frac{32}{100} \times 25 = \frac{25}{100} \times 32 = \frac{1}{4} \times 32 = 8$.

d) $\frac{900}{100} \times 8.1 = 9 \times 8.1 = 72.9$.

Important: From part c), you can see that 32% of 25 is the same as 25% of 32. This is always true. **x% of y is equivalent to y% of x**.

Example 6:   Find the percent change.

a) from 9 to 17   b) from 35 to 49   c) from 132 to 120

Solution: a) To find the percent change, you can use the following formula:

$$\frac{\text{New Value} - \text{Old Value}}{\text{Old Value}}$$

(Remember: **NOO**...)

A common mistake students make is using $\frac{\text{New} - \text{Old}}{\text{New}}$.

That is why it is important to remember the acronym "NOO."

From 9 to 17, the original value is 9. So we have,

$$\frac{17-9}{9} = \frac{8}{9} = 0.8888... = 88.9\% \text{ increased}$$

b) From 35 to 49, the original value is 35. So we have,

$$\frac{49-35}{35} = \frac{14}{35} = \frac{2}{5} = 0.4 = 40\% \text{ increased}$$

c) From 132 to 120, the original value is 132. So we have,

$$\frac{120-132}{132} = \frac{-12}{11 \times 12} = \frac{-1}{11} = -0.0909... = 9.09\% \text{ decreased.}$$

20/20 Math

Practice 6

Use mental math to calculate each quotient.

1a) $83 \div 9$ 1b) $919 \div 9$ 1c) $1010 \div 90$

2a) $72 \div 11$ 2b) $29 \div 8$ 2c) $220 \div 60$

Find the unit rate.

3) $\dfrac{540 \text{ feet}}{20 \text{ seconds}}$ 4) $\dfrac{45 \text{ pounds}}{25 \text{ dollars}}$ 5) $\dfrac{729 \text{ students}}{3 \text{ teachers}}$

Calculate the percent.

6) $\dfrac{17}{20}$ 7) $\dfrac{39}{50}$ 8) $\dfrac{124}{300}$

9) $\dfrac{3}{8}$ 10) $\dfrac{21}{30}$ 11) $\dfrac{35}{63}$

Convert each percent to decimal or each decimal to percent.

12) 67%   13) 110%   14) 2.15

15) 0.24%   16) 0.7   17) 1240%

Find each of the following.

18) 60% of 120   19) 15% of 15   20) 200% of 51

Bonus: Find the percent change: From 36 to 3600.

34

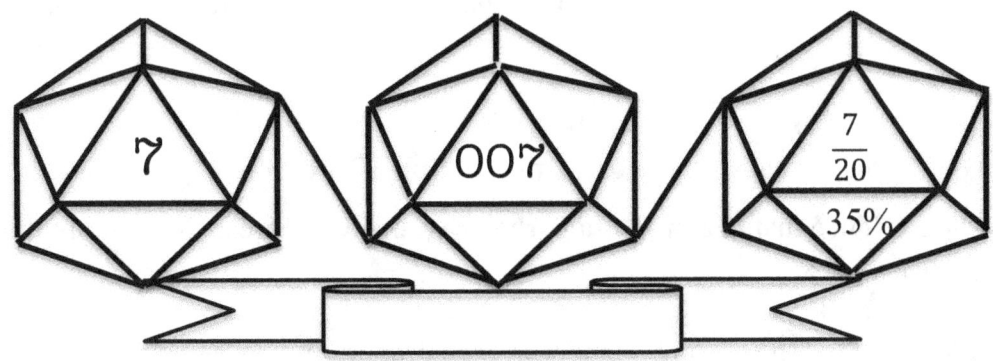

Lesson 7: Ratios, Proportions, and More Percentages

In the pevious lesson, you learned how to calculate the unit rate. You saw that a **rate** is a comparison of two quantities that have **different units**. For example, if 10 pounds of apples cost $20, then the rate is 20 dollars for 10 pounds. The unit rate is 2 dollars for 1 pound.

A **ratio** is a comparison of two quantities that have the **same unit**. For example, if a class has 14 girls and 11 boys, then **the ratio of girls to boys** can be expressed as 14 to 11, or 14:11, or $\frac{14}{11}$. Similarly, the ratio of girls to the total number of students in the class is 14 to 25, or 14:25, or $\frac{14}{25}$.

Example 1: a) A poster is 2 ft long and 8 in wide. Find the ratio of its length to its width.

b) Last season, the Lakers won 56 games and lost 26. Find the ratio of the number of games won to the total number of games.

Solution: a) Note that it is WRONG to say the ratio is 2 to 8 or 1 to 4, because that would mean the width is four times as long as the length. But, the length is 2 feet, which equals 24 inches. It is longer than the width. So, the ratio should be 24 to 8 or 3 to 1. As you can see, you have to make the units the same before you can compare them. Otherwise, the answer does not make sense.

20/20 Math

b) Since the Lakers won 56 games out of 82 games played, the ratio is $\frac{56}{82} = \frac{28}{41}$.

When two ratios are equal, they form a **proportion**. For example, $\frac{1}{2} = \frac{5}{10}$ is a proportion. Notice that in a proportion, **the cross products** are also equal. In this case, $1 \times 10 = 2 \times 5$.

***Example 2*:**   Find the missing number in each proportion.

a) $\frac{\square}{4} = \frac{3}{5}$ b) $\frac{8}{9} = \frac{2}{\square}$ c) $\frac{11}{100} = \frac{\square}{11}$

***Solution*:**   a) Since the cross products must equal, $5 \times \square = 4 \times 3 = 12$. That means the missing number is $12 \div 5 = \mathbf{2.4}$.

b) $8 \times \square = 2 \times 9 = 18$.
The missing number is $18 \div 8 = 9 \div 4 = \mathbf{2.25}$.

c) $11 \times 11 = 100 \times \square$. The missing number is $121 \div 100 = \mathbf{1.21}$.

Important: To answer the more challenging percentage problems, you can use the following proportion:

$$\frac{part}{whole} = \frac{percent}{100}.$$

***Example 3*:**   a) 60% of what number is 48?

b) What percent is 5 out of 8?

c) What number is 32% of 11?

d) 78 is 120% of what number?

Important: The phrases x "**percent of**" y or x "**out of**" y mean **y is the whole**.

***Solution*:**   a) 60 "percent of" what number means "what number" is the whole. In other words, the **whole** is the unknown. So, 48 must be

the **part**. Hence, the proportion is $\frac{48}{whole} = \frac{60}{100}$.
Setting the cross products equal, we have 60 × whole = 4800.
So, the **whole = 4800 ÷ 60 = 80**.

b) 8 is the whole, and 5 is the part. The percent is unknown.
Hence, the proportion is $\frac{5}{8} = \frac{percent}{100}$.
Setting the coss products equal, we have 8 × percent = 5 × 100.
So, percent = $\frac{5}{8}$ × 100 = 5 × 0.125 × 100 = 0.625 × 100 = **62.5%**.

c) Percent is 32. The whole is 11. The part is unknown.
Hence, the proportion is $\frac{part}{11} = \frac{32}{100}$.
Setting the cross products equal, we have 100 × part = 11 × 32.
So, part = 352 ÷ 100 = **3.52**.

d) Percent is 120. The whole is unknown. The part is 78.
Hence, the proportion is $\frac{78}{whole} = \frac{120}{100}$.
Setting the cross products equal, we have 120 × whole = 78 × 100.
So, whole = $\frac{78 \times 100}{120}$ = 78 × $\frac{5}{6}$ = $\frac{78}{6}$ × 5 = 13 × 5 = **65**.

Example 4: Mrs. Anderson went out to eat at a restaurant with her two friends. If the bill was $52.05, and they gave the waitress a 20% tip, how much did they pay altogether?

Solution: Method 1: 52.05 + (20% of 52.05) = 52.05 + $\frac{20}{100}$ × 52.05 =
52.05 + $\frac{1}{5}$ × 52.05 = 52.05 + 10.41 = **$62.46**.

Method 2: Paying 20% tip means they paid 120% of the bill. So,
$\frac{120}{100}$ × 52.05 = 1.2 × 52.05 = **$62.46**.

20/20 Math

Practice 7

1) A rectangle is 16 cm long and 12 cm wide. Find the ratio of the width to the perimeter of the rectangle.

2) Ms. Wang's class currently has 35 students, of which 19 are boys. If one boy moves to another school tomorrow, find the ratio of the remaining boys to girls in Ms. Wang's class.

Find the missing number in each proportion.

3) $\dfrac{20}{21} = \dfrac{6}{\square}$ 	4) $\dfrac{14}{\square} = \dfrac{3.5}{13}$ 	5) $\dfrac{26}{91} = \dfrac{\square}{7}$

6) $\dfrac{\square}{111} = \dfrac{3}{0.5}$ 	7) $\dfrac{-15}{23} = \dfrac{3}{\square}$ 	8) $\dfrac{30}{\square} = \dfrac{2.5}{-4}$

9) 45% of what number is 20.25? 	10) What number is 95% of 95?

11) What percent is 49 out of 90? 	12) 110 is 110% of what number?

13) 300% of what number is 12.6? 	14) What number is 0.8% of 0.7?

15) 0.056 out of 0.08 is what percent? 	16) 0.25% of what number is 0.9?

17) Because of extra credits, Chris got 27 out of 25 on his last test. What is the percentage Chris received?

18) Kobe Bryant can make 84% of his free throws. If he attempted a total of 75 free throws in his last 5 games, how many free throws did Kobe make?

19) Last month, David spent $1575. If this represents 35% of his income, how much money did David make last month?

20) Ashley bought a laptop at Best Buy with a 15% discount, and she has to pay a 10% tax on the discounted price. What is Ashley's final amount if the original price was $950?

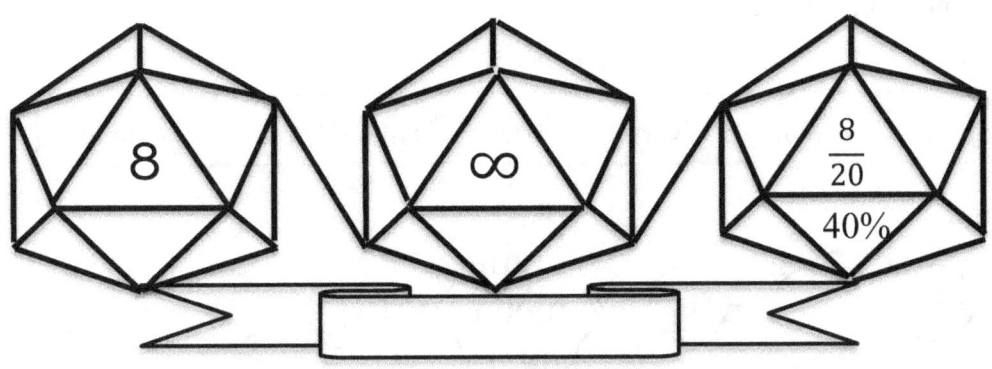

Lesson 8: Metric Units Conversion

In lesson 1, you learned how to quickly multiply or divide by power of tens. In this lesson, you will use those techniques to convert units in the **metric system**. This is a decimal-based system, which means its units are related to one another by factors of 10. For example, the standard unit for **length** in the metric system is the meter, and 1 m = 10^2 **centi**meter = 100 cm. Equivalently, 1 cm = 0.01 m. Likewise, 1 m = 10^3 **milli**meter = 1000 mm. So, 1 mm = 0.001 m.

As you can see, to convert units in the metric system, you first need to know what the prefixes stand for. Here is the table of the common prefixes.

Prefixes	Kilo	Hecto	Deca	Base Units	Deci	Centi	Milli
Abbreviation	k	h	da	gram (g) meter (m) liter (L)	d	c	m
Factor	1000	100	10		0.1	0.01	0.001

In addition to the base unit of **length**, you need to know that the base units of **weight** is gram, and **capacity** is liter.

For example, based on the table, 1 kg = 1000 g. 1 kL = 1000 L. Similarly, 1 mg = 0.0001 g, which means 1g = 1000 mg. Although you can use the above table to convert metric units, there is an easier way.

39

First, you should memorize this mnemonic:

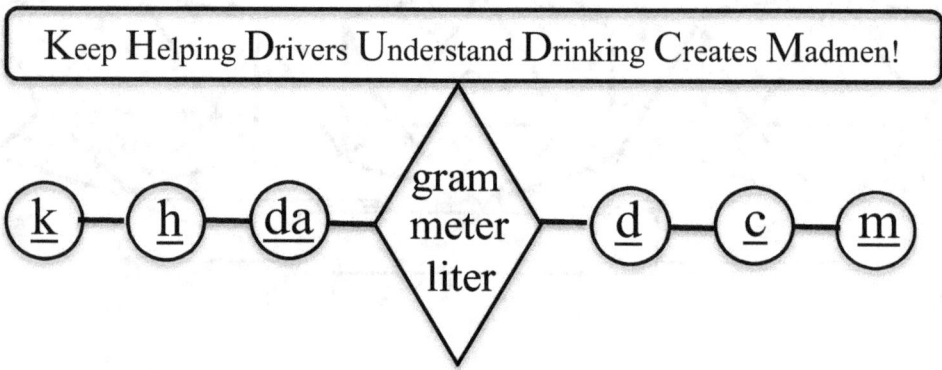

Then, to do a conversion, you should put down the **given** under the correct unit and move the decimal point until you get to the desired unit.

Example 1:   Convert each of the following:

   a) 3.45 kL = __ mL   b) 4 cg = __ mg   c) 0.09 m = __ cm

Solution: a) Here we are given 3.45 **kiloliter**. Put it under "Keep."

Keep Helping Drivers Understand Drinking Creates Madmen

3.45 kL   mL

We want to get to **milliliter**. So, we need to move the decimal 6 times to the **right** (i.e., from "Keep" to "Madmen").
So, 3.45 kL = 3,450,000.0 mL = 3,450,000 mL.

b) Here we are given 4 or 4.0 **centigram**. Put it under "Creates."

Keep Helping Drivers Understand Drinking Creates Madmen

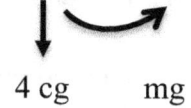

4 cg   mg

We want to get to **milligram**. So, we need to move the decimal 1 time to the **right** (i.e., from "Creates" to "Madmen").
So, 4 cg = 4.0 cg = 40 mg.

c) Here we are given 0.09 **meter**. That is the base unit. So, put it under "Understand."

Keep Helping Drivers Understand Drinking Creates Madmen

0.09 m   cm

We want to get to **centimeter**. So, we need to move the decimal 2 times to the **right** (i.e., from "Understand" to "Creates").
So, 0.09 m = 9 cm.

Example 2:   Convert each of the following:

a) 13.5 mm = __ dm

b) 46532 mg = __ kg

c) 0.000167 mL = __ L

Solution: a) Here we are given 13.5 **millimeter**. Put it under "Madmen."

Keep Helping Drivers Understand Drinking Creates Madmen

dm   13.5 mm

We want to get to **decimeter**. So, we need to move the decimal 2 times to the **left** (i.e., from "Madmen" to "Drinking").
So, 13.5 mm = 0.135 dm.

b) Here we are given 46532.0 **milligram**. Put it under "Madmen."

Keep Helping Drivers Understand Drinking Creates Madmen

kg   46532 mg

We want to get to **kilogram**. So, we need to move the decimal 6 times to the **left** (i.e., from "Madmen" to "Keep").
So, 46532.0 mg = 0.046532 kg.

c) Here we are given 0.000167 **milliliter**. So, put it under "Madmen."

Keep Helping Drivers Understand Drinking Creates Madmen

       L   0.000167 mL

We want to get to **liter**, which is the base unit. So, we need to move the decimal 3 times to the **left** (i.e., from "Madmen" to "Understand"). So, 0.000167 mL = 0.000,000,167 L.

Example 3: From point A, an airplane flies 4110 km, and arrives at point B 6 hours later. What is the average speed of the plane, in meters per second?

Solution: Average speed or rate = distance ÷ time.

The distance is 4110 km. To go to meters, put it under "Keep," and move the decimal 3 times to the **right**.

Keep Helping Drivers Understand Drinking Creates Madmen

4110 km   m

So, 4110 km = 4,110,000 m.

The time is 6 hours, which equals 6 × 60 = 360 minutes.
360 minutes also equals 360 × 60 = 21600 seconds.

Therefore, the avearge speed is
  4,110,000 m ÷ 21600 seconds = 190 meter per second.

Another method:
Average speed = 4110 km ÷ 6 hours = 685 kilometer per hour.

But, 685 km = 685,000 m, and 1 hour = 3600 seconds.

Therefore, the average speed is
  685,000 m ÷ 3600 seconds = 190 meter per second.

Example 4: Suppose a rectangel is 5 meters long and 4 meters wide. Find the area of the rectangle in square centimeter.

Solution: The length is 5 m. To go to centimeter, put it under "Understand," and move the decimal 2 times to the **right**.

Keep Helping Drivers Understand Drinking Creates Madmen

             5 m   cm

So, 5 m = 500 cm. Similarly, 4 m = 400 cm.

Since the area of the rectangle is length × width, we have
    500 cm × 400 cm = **200,000 square centimeter**.

Common Mistake:

Since the area of the rectangle is length × width, we have
    5 m × 4 m = 20 square meter.

We know that 1 m = 100 cm.
Therefore, 20 m² = 2000 cm². This is WRONG.

Even though 1 m = 100 cm, 1 m² is NOT 100 cm².

In fact, 1 m² = 1 m × 1 m = 100 cm × 100 cm = 10,000 cm².

Practice 8

Convert each of the following.

1) 37 L = ___ mL

2) 1.0023 g = ___ kg

3) 0.068 km = ___ mm

4) 1,000,000 cm = ___ km

5) 0.057 L = ___ kL

6) 350,000 dm = ___ mm

7) $\frac{1}{3}$ kg = ___ cg

8) $\frac{2}{9}$ g = ___ mg

9) $\frac{1}{1000}$ cL = ___ dL

10) 0.00382 dag = ___ g

Order from least to greatest.

11) .245 km, 650 cm, 312 m, 50,000 mm

12) 0.64g, 0.0075 kg, 80 cg, 7400 mg

13) 0.1 kL, 20 L, 500 cL, 40,000 mL

14) Find the perimeter of the rectangle in meter.

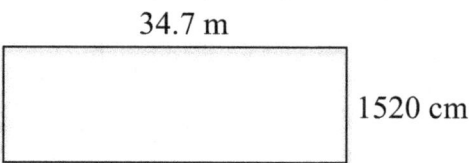

15) Find the area of the square in square centimeter.

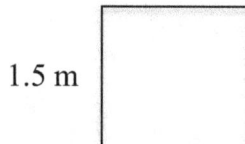

16) Find the perimeter of the triangle in kilometer.

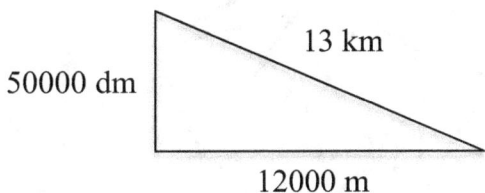

17) Convert 48 kilograms per year to gram per month.

18) Convert 1250 liters per week to kiloliter per year. (Hint: 1 year = 52 weeks).

19) Express the following sum in gram: 0.0592 kg + 2460 mg + 12.4 g.

20) A 2.3 liter drink costs $3.91. Express the cost in cent per milliliter.

Bonus: Convert 4.5 m³ to cm³.

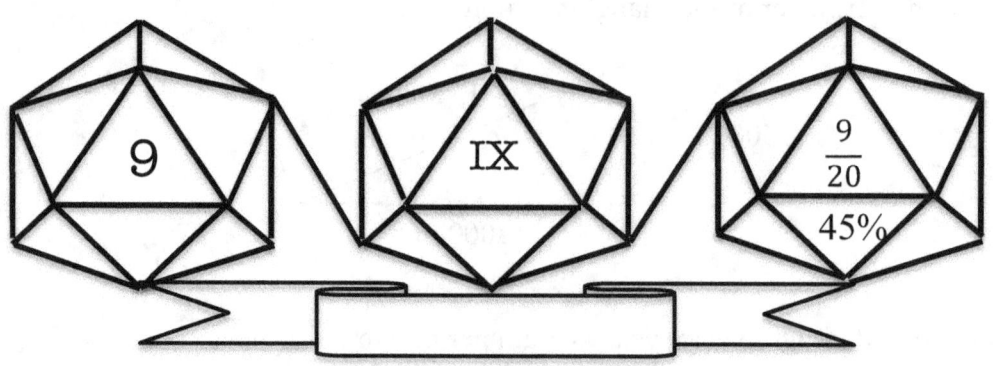

Lesson 9: Customary Units Conversion

Unlike metric units, customary units are not related to one another by factors of tens. For example, 1 foot = 12 inches, which means 1 inch = $\frac{1}{12}$ feet. In fact, you should use a calculator if you need a decimal approximation of the answer. Here are the conversion factors for all the customary units of length.

Length
1 foot (ft) = 12 inches (in.)
1 yard (yd) = 3 feet = 36 inches
1 mile (mi) = 5280 feet = 1760 yards

Here are the steps to do a conversion.

1) Put down the **given**.

2) Multiply the given by the conversion factor $\frac{\text{desired unit}}{\text{given unit}}$.

3) Cancel out the given unit and simplify.

Pre-Algebra

Example 1: Convert each of the following:

a) 6 ft = __ in.   b) 100 in. = __ ft   c) 4.5 yd = __ in.

d) 2 mi = __ ft   e) 56 ft = __ yd

Solution: a) Here we are given 6 feet. We want to get to inches. Since 1 foot = 12 inches, we multiply 6 feet by $\frac{12 \text{ in.}}{1 \text{ ft}}$.
Hence, 6 ft × $\frac{12 \text{ in.}}{1 \text{ ft}}$ = 6 × 12 in. = 72 in.

b) Here we are given 100 inches. We want to get to feet. Since 1 foot = 12 inches, we multiply 100 inches by $\frac{1 \text{ ft}}{12 \text{ in.}}$.
Hence, 100 in. × $\frac{1 \text{ ft}}{12 \text{ in.}}$ = 100 ft ÷ 12 = $8\frac{1}{3}$ ft.

Notice that in this problem, we use the same fact as part (a). That is, 1 foot = 12 inches. However, the conversion factor is different, it is $\frac{1 \text{ ft}}{12 \text{ in.}}$ versus $\frac{12 \text{ in.}}{1 \text{ ft}}$. So, in the end we have to divide 100 by 12 instead of multiplying. In fact, that is the main benefit of solving these problems this way. You do not have to determine whether you should multiply or divide from the beginning. This technique will lead you to the right path at the end.

c) Here we are given 4.5 yard. We want to get to inches. Since 1 yard = 36 inches, we multiply 4.5 yard by $\frac{36 \text{ in.}}{1 \text{ yd}}$.
Hence, 4.5 yd × $\frac{36 \text{ in.}}{1 \text{ yd}}$ = 4.5 × 36 in. = 162 in.

d) Here we are given 2 miles. We want to get to feet. Since 1 mile = 5280 feet, we multiply 2 miles by $\frac{5280 \text{ ft}}{1 \text{ mi}}$.
Hence, 2 mi × $\frac{5280 \text{ ft}}{1 \text{ mi}}$ = 2 × 5280 ft = 10560 ft.

e) Here we are given 56 teet. We want to get to yard. Since 1 yard = 3 feet, we multiply 56 feet by $\frac{1 \text{ yd}}{3 \text{ ft}}$.
Hence, 56 ft × $\frac{1 \text{ yd}}{3 \text{ ft}}$ = 56 yd ÷ 3 = $18\frac{2}{3}$ yd.

Here are the conversion factors for the customary units of weight.

Weight
1 pound (lb) = 16 ounces (oz)
1 ton (T) = 2000 pounds

Example 2: Convert each of the following:

a) 12.5 lb = ___ oz

b) 256 oz = ___ lb

c) 7800 lb = ___ T

d) $\frac{3}{8}$ T = ___ oz

Solution: a) Here we are given 12.5 pounds. We want to get to ounces. Since 1 pound = 16 ounces, we multiply 12.5 lb by $\frac{16 \text{ oz}}{1 \text{ lb}}$.
Hence, $12.5 \text{ lb} \times \frac{16 \text{ oz}}{1 \text{ lb}} = 12.5 \times 16 \text{ oz} = 200 \text{ oz}$.

b) Here we are given 256 ounces. We want to get to pounds. Since 1 pound = 16 ounces, we multiply 256 ounces by $\frac{1 \text{ lb}}{16 \text{ oz}}$.
Hence, $256 \text{ oz} \times \frac{1 \text{ lb}}{16 \text{ oz}} = 256 \text{ lb} \div 16 = 16 \text{ lb}$.

c) Here we are given 7800 pounds. We want to get to tons. Since 1 ton = 2000 pounds, we multiply 7800 lb by $\frac{1 \text{ ton}}{2000 \text{ lb}}$.
Hence, $7800 \text{ lb} \times \frac{1 \text{ ton}}{2000 \text{ lb}} = 7800 \text{ tons} \div 2000 = 3.9 \text{ tons}$.

d) Here we are given $\frac{3}{8}$ ton. We want to get to ounces.
First, use 1 ton = 2000 pounds to get to pounds.
Then use 1 pound = 16 ounces to finally get to the desired unit.
Hence, $\frac{3}{8} \text{ T} \times \frac{2000 \text{ lb}}{1 \text{ T}} \times \frac{16 \text{ oz}}{1 \text{ lb}} = \frac{3}{8} \times 2000 \times 16 \text{ oz} = 12000 \text{ oz}$.

Pre-Algebra

For capacity, draw the following diagram to help you remember the conversion factors. It is easier to remember this diagram than memorize the table.

20/20 Math

Here is the table version. Note that the last row is not in the diagram.

Capacity
1 gallon (gal) = 4 quarts (qt)
1 quart = 2 pints (pt)
1 pint = 2 cups (c)
1 cup = 8 fluid ounces (fl oz)

Example 3:   Convert each of the following:

   a) 16.5 gal = __ qt   b) 92 pt = __ qt

   c) 96 c = __ gal   d) 7.1 c = __ fl oz

Solution: a) Here we are given 16.5 gallons. We want to get to quarts. Since 1 gallon = 4 quarts, we multiply 16.5 gal by $\frac{4 \text{ qt}}{1 \text{ gal}}$.
Hence, 16.5 gal × $\frac{4 \text{ qt}}{1 \text{ gal}}$ = 16.5 × 4 qt = 66 qt.

b) Here we are given 92 pints. We want to get to quarts. Since 1 quart = 2 pints, we multiply 92 pt by $\frac{1 \text{ qt}}{2 \text{ pt}}$.
Hence, 92 pt × $\frac{1 \text{ qt}}{2 \text{ pt}}$ = 92 qt ÷ 2 = 46 qt.

c) Here we are given 96 cups. We want to get to gallons. From the diagram, 1 gal = 16 cups, we multiply 96 c by $\frac{1 \text{ gal}}{16 \text{ c}}$.
Hence, 96 c × $\frac{1 \text{ gal}}{16 \text{ c}}$ = 96 gal ÷ 16 = 6 gal.

Notice that in this problem, it is easier to use the diagram to get 1 gal = 16 cups.

d) Here we are given 7.1 cups. We want to get to fluid ounces. Since 1 cup = 8 fluid ounces, we multiply 7.1 c by $\frac{8 \text{ fl oz}}{1 \text{ c}}$.
Hence, 7.1 c × $\frac{8 \text{ fl oz}}{1 \text{ c}}$ = 7.1 × 8 fl oz = 56.8 fl oz.

Practice 9

Convert each of the following.

1) 4 yd = ___ in.

2) 12 ft = ___ in.

3) 5.1 mi = ___ yd

4) 3 yd = ___ ft

5) 12672 ft = mi ___

6) 500 in. = ___ yd

7) $\frac{13}{25}$ T = ___ lb

8) $\frac{3}{8}$ lb = ___ oz

9) 38400 oz = ___ T

10) 0.099 T = ___ oz

11) 20 fl oz = ___ c

12) 10.2 gal = ___ c

13) 24 qt = ___ pt

14) 36 pt = ___ gal

15) 19 pt = ___ c

16) 3 gal = ___ fl oz

17) 56 c = ___ qt

18) 512 fl oz = ___ gal

19) 2.02 gal = ___ c

20) 70.4 c = ___ pt

Bonus: Convert 9 gal to fl oz.

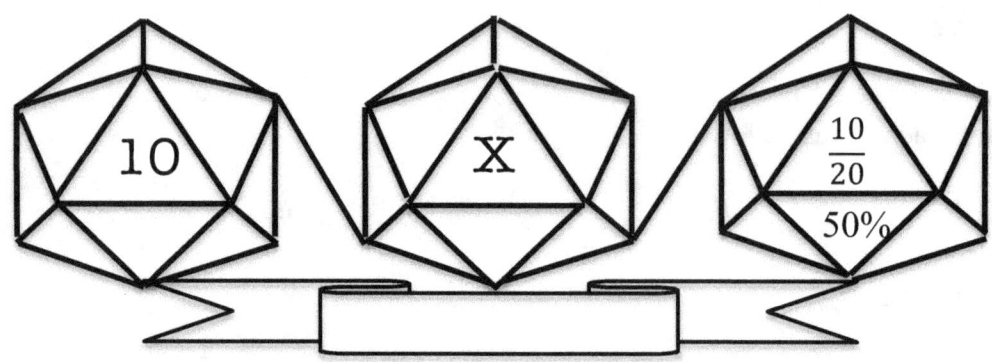

Lesson 10: Variables and Expressions

In primary school math, you learn to work with numbers. In algebra, you learn to reason with variables. A **variable** is a letter that hold the value of a number. For example, we can let the letter x hold the value of the missing number in a proportion such as $\frac{1}{2} = \frac{6}{x}$. That way, finding the missing number is the same as solving for x.

Now, a **numerical expression** is a math sentence consists of numbers and operations. For example, $5 + 8 \div 2$ is a numerical expression.

An **algebraic expression** has at least one variable. For example, $x^2 - x + 10$ is an algebraic expression. Notice that the value of this expression depends on what x is equal to. Suppose we know $x = 3$, then we can **evaluate** or calculate this expression by replacing x with 3. Hence, $x^2 - x + 10 = 3^2 - 3 + 10 = 9 - 3 + 10 = 16$. On the other hands, if $x = 5$, then $x^2 - x + 10 = 5^2 - 5 + 10 = 30$.

Example 1: Evaluate each expression.

a) $-2x + 4y$ for $x = 6$ and $y = -7$ b) $\frac{a+2b}{8}$ for $a = 4$ and $b = 6$

c) $9(2x + 14)$ for $x = 3.5$ d) $x^4 + y^2$ for $x = 3$ and $y = 9$

Solution: a) $-2x$ means "-2 times x," and $4y$ means "4 times y."
So, $-2x + 4y = -2(6) + 4(-7) = -12 + (-28) = -40$.

b) $\dfrac{a+2b}{8} = \dfrac{4+2(6)}{8} = \dfrac{4+12}{8} = \dfrac{16}{8} = 2.$

c) $9(2x + 14) = 9(2(3.5) + 14) = 9(7 + 14) = 9(21) = 189.$

d) $x^4 + y^2 = (3)^4 + (9)^2 = 81 + 81 = 162.$

Important: In an algebraic expression, plus or minus signs separate each **term**.

For example, $x^2 + 5x - 12$ contains three terms. They are x^2, $5x$, and -12.

Notice that we count "5 times x" as one term because only plus or minus separates a term, multiplication does not. In fact, the number in front of the variable is called the **coefficient**.

When two or more terms have the same variable(s) and exponent, they are **like terms**. For example, in the expression $3x^2 + 2x + 5x^2 - 7x + 4$, $3x^2$ and $5x^2$ are like terms; $2x$ and $-7x$ are also like terms.

To simplify an algebraic expression, you can combine like terms by adding their coefficients. So, $3x^2 + 2x + 5x^2 - 7x + 4 = 8x^2 - 5x + 4$.

Example 2: Simplify each expression.

 a) $-4x^3 + 2y^2 + 2x^3 + 3x^2$ b) $4m + 5n - 6m$

 c) $9xyz + 10xy + 10yz + 2xyz$ d) $t^4 + t^2 + t + 1$

Solution:
a) $-4x^3$ and $2x^3$ are like terms. So, you can combine them into $(-4 + 2)x^3 = -2x^3$. The rest of the terms are different.
Hence, $-4x^3 + 2y^2 + 2x^3 + 3x^2 = -2x^3 + 3x^2 + 2y^2$

b) $4m + 5n - 6m = -2m + 5n$

c) $9xyz$ and $2xyz$ are like terms.
The remaining terms are different.
Hence, $9xyz + 10xy + 10yz + 2xyz = 11xyz + 10xy + 10yz$

d) None of the terms in this problem are like terms because they all have different exponents. Hence, the expression is already simplified.

Based on order of operations, when you have $3 \times (4 + 5)$, you are supposed to do $4 + 5 = 9$ first. Then you do $3 \times 9 = 27$. However, if you do $3 \times 4 + 3 \times 5$, you would get the same result because $12 + 15 = 27$. So, $3 \times (4 + 5) = 3 \times 4 + 3 \times 5$.

Important: $a(b + c) = ab + ac$. This is called the **Distributive Property**.

The distributive property is a very useful tool when it comes to mental math.

Example 3: Use mental math to calculate each expression.

        a) 21×3.4 b) 19^2 c) 15×42

Solution: a) You can break down 3.4 into $3 + 0.4$.
So, $21 \times 3.4 = 21 \times (3 + 0.4) = 21 \times 3 + 21 \times 0.4 = 63 + 8.4 = 71.4$.

    b) $19^2 = 19 \times 19 = 19 \times (20 - 1) = 19 \times 20 - 19 \times 1$
                                          $= 380 - 19 = 361$.

    c) $15 \times 42 = 15 \times (40 + 2) = 15 \times 40 + 15 \times 2 = 600 + 30 = 630$.

Example 4: Use distributive property to simplify each expression.

    a) $2(3x^2 + 5x - 6) + 4(x^2 - 8)$

    b) $12m^3 + 4m^2 - 6m(m^2 - m - 2)$

    c) $(2a + 3b) \times 4 + 13a - 15b$

Solution: a) $2(3x^2 + 5x - 6) = 6x^2 + 10x - 12$.
$4(x^2 - 8) = 4x^2 - 32$.
Hence, $2(3x^2 + 5x - 6) + 4(x^2 - 8) = 6x^2 + 10x - 12 + 4x^2 - 32$
                                                          $= 10x^2 + 10x - 44$.

    b) $-6m(m^2 - m - 2) = -6m^3 + 6m^2 + 12m$.
Hence, $12m^3 + 4m^2 - 6m(m^2 - m - 2) =$
$12m^3 + 4m^2 - 6m^3 + 6m^2 + 12m = 6m^3 + 10m^2 + 12m$.

    c) $(2a + 3b) \times 4 = 4 \times (2a + 3b) = 8a + 12b$.
Hence, $(2a + 3b) \times 4 + 13a - 15b =$
$8a + 12b + 13a - 15b = 21a - 3b$.

Practice 10

Evaluate each expression.

1) x^2 if $x = 11.5$

2) $3x^2 - 5x + 7$ if $x = 10$

3) $31(x^3 + x)$ if $x = 3$

4) $374 \div x \times y$ if $x = 11$ and $y = 20$

5) $x^4y - 499$ if $x = 5$ and $y = 4$

6) $3.5x - 0.6$ if $x = 35$

7) $(2p + q)^2$ if $p = -9$ and $q = 6$

8) $\dfrac{1}{k} - \dfrac{1}{2k} + \dfrac{1}{3k}$ if $k = 3$

Simplify each expression.

9) $4 + 2x^2 - 6x + 15x^2 - 4$

10) $-a^2b - ab^2 - 2ab - 2ab^2 - 3a - 3b$

11) $10mn + m + n + mn$

12) $1 + xyz - 2 - \dfrac{1}{3}(xyz)$

13) $-(2x - 3y) + 13x - 13y$

14) $20(15x - 7) - 2(x^2 - 13)$

15) $2 - 2(2a - 2b - 2) - 6$

16) $m - (m^2 + m + 1) \times 5$

Use mental math to calculate each expression.

17) 108×12

18) 21^2

19) 16×25.3

20) 212×1.05

Bonus: Given $x = -1$, evaluate $x^2 - x^3 - x^4 - x^5 - x^6 - x^7 - x^8 - x^9$.

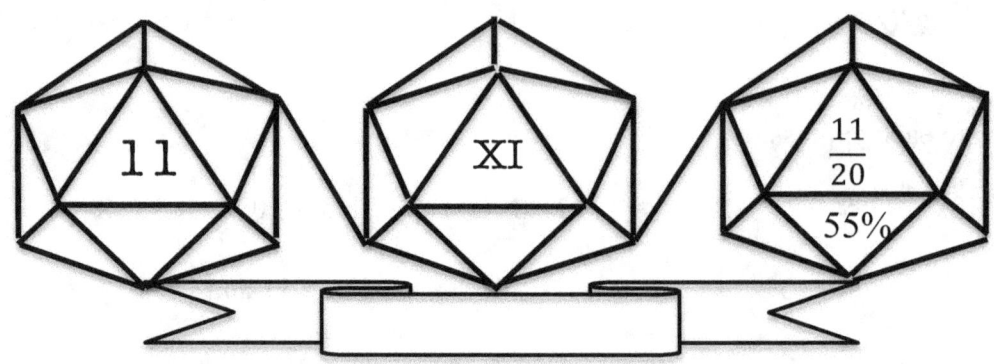

Lesson 11: Solving Linear Equations

Setting two algebraic expressions equal to each other produces an **equation**. For example, $2x - 5 = x + 7$ is an equation, NOT an expression, because it contains the equal sign. Solving an equation means finding the value(s) of the variable that will make the equation true. You will learn how to solve different types of equations throughout the study of algebra. Solving linear equations is the focus of this lesson.

An equation is **linear** if the highest exponent of the variables is 1. For example, $x^2 + 7x + 10 = 0$ is non-linear because the highest exponent is 2. You already learned how to solve simple linear equations such as $\square + 5 = 7$, since first grade. Back then, you solved such a problem by inspection, or using mental math. However, if an equation gets more complex, it will be difficult to figure out the solution just by looking at it. Instead, you can use the following properties to help you move from one equation to another simpler and equivalent equation.

The Addition Property of Equality: If $a = b$, then $a + c = b + c$.

The Subtraction Property of Equality: If $a = b$, then $a - c = b - c$.

The Multiplication Property of Equality: If $a = b$ and $c \neq 0$, then $ac = bc$.

The Division Property of Equality: If $a = b$ and $c \neq 0$, then $a \div c = b \div c$.

Example 1: Solve.

    a) $x + 52 = -23$ b) $-34 = 17 - m$

    c) $2y + 19 = 46$ d) $\dfrac{x}{8} - 1.28 = 3.72$

Solution:

a) $x + 52 = -23$

$x + 52 - 52 = -23 - 52$	Subtract 52 on both sides
$\boxed{x = -75}$	Simplify

b) $-34 = 17 - m$

$-34 + m = 17 - m + m$	Add m on both sides
$-34 + m = 17$	Simplify
$-34 + 34 + m = 17 + 34$	Add 34 on both sides
$\boxed{m = 51}$	Simplify

c) $2y + 19 = 46$

$2y + 19 - 19 = 46 - 19$	Subtract 19 on both sides
$2y = 27$	Simplify
$\boxed{y = 13.5}$	Divide by 2 on both sides

d) $\dfrac{x}{8} - 1.28 = 3.72$

$\dfrac{x}{8} - 1.28 + 1.28 = 3.72 + 1.28$	Add 1.28 on both sides
$\dfrac{x}{8} = 5$	Simplify
$\boxed{x = 40}$	Multiply 8 on both sides

Example 2: Solve.

a) $\dfrac{-x+3}{50} = \dfrac{-2}{15}$ b) $\dfrac{1}{4}x - \dfrac{1}{5} = -x + \dfrac{3}{5}$

c) $-(2x - 7) = 3(-x + 5) - 10$ d) $1.3(x + 2) + x = 2x - 3.2$

Solution:

a) $\dfrac{-x+3}{50} = \dfrac{-2}{15}$

$-15x + 45 = -100$ Cross multiply

$-15x = -145$ Subtract 45 on both sides

$\boxed{x = \dfrac{145}{15} = \dfrac{29}{3} = 9.666\ldots}$ Divide by -15 on both sides

b) $\dfrac{1}{4}x - \dfrac{1}{5} = -x + \dfrac{3}{5}$

$\dfrac{5}{4}x - \dfrac{1}{5} = \dfrac{3}{5}$ Add x on both sides

$\dfrac{5}{4}x = \dfrac{4}{5}$ Add $\dfrac{1}{5}$ on both sides

$\boxed{x = \dfrac{16}{25}}$ Multiply by $\dfrac{4}{5}$ on both sides

c) $-(2x - 7) = 3(-x + 5) - 10$

$-2x + 7 = -3x + 15 - 10$ Distribute

$x + 7 = 15 - 10$ Add 3x on both sides

$x + 7 - 7 = 15 - 10 - 7$ Subtract 7 on both sides

$\boxed{x = -2}$ Simplify

d) $1.3(x + 2) + x = 2x - 3.2$

$1.3x + 2.6 + x = 2x - 3.2$ Distribute

$2.3x + 2.6 = 2x - 3.2$ Combine like terms

$$2.3x - 2x + 2.6 = 2x - 2x - 3.2 \qquad \text{Subtract 2x on both sides}$$

$$0.3x + 2.6 = -3.2 \qquad \text{Simplify}$$

$$0.3x + 2.6 - 2.6 = -3.2 - 2.6 \qquad \text{Subtract 2.6 on both sides}$$

$$0.3x = -5.8 \qquad \text{Simplify}$$

$$\boxed{x = \frac{-5.8}{0.3} = \frac{-58}{3} = -19.333\ldots} \qquad \text{Divide by 0.3 on both sides}$$

Example 3: The measures of the three sides of a triangle are $x + 8$, $4x + 15$ and $10 - 2x$. If the perimeter of the triangle is 39. Find the value of x.

Solution: Add up the three sides of the triangle and set it equals to 39, we have, $x + 8 + 4x + 15 + 10 - 2x = 39$.

$\Rightarrow \quad 3x + 33 = 39 \qquad$ Combine like terms
$\Rightarrow \quad 3x + 33 - 33 = 39 - 33 \qquad$ Subtract 33 on both sides
$\Rightarrow \quad 3x = 6 \qquad$ Simplify
$\Rightarrow \quad x = 2 \qquad$ Divide by 3 on both sides

Example 4: Paul's math scores on his first three tests are 83, 94 and 87. What must Paul receive on his fourth test in order to have an average of 90 or better in the class?

Solution: Let x be the number of points Paul will get on his fourth test.
To find the average grade, add up the 4 scores and divide by 4.

Thus, we have $\frac{83+94+87+x}{4} = 90$.

$\Rightarrow \quad 83 + 94 + 87 + x = 360 \qquad$ Multiply 4 on both sides
$\Rightarrow \quad 264 + x = 360 \qquad$ Simplify
$\Rightarrow \quad 264 - 264 + x = 360 - 264 \qquad$ Subtract 264 on both sides
$\Rightarrow \quad x = 96 \qquad$ Simplify

Thus, Paul must score 96 points or higher on his fourth test to have an average of 90 or better.

Practice 11

Solve.

1) $7p + 9 = 37$

2) $1 - x = 10 + 2x$

3) $3(x - 5) = x - 15$

4) $\dfrac{1}{2} = -3.5 + 7x$

5) $\dfrac{2m}{11} - 8 = 4$

6) $\dfrac{x}{3} + \dfrac{x}{4} - 5 = \dfrac{5x}{12} + 8$

7) $1.3y - 5(y + 2) = 27$

8) $-(2 - 2x) + 3 = 0.5x + 5.5$

9) $\dfrac{a - 20}{44} = \dfrac{4}{-5}$

10) $\dfrac{2}{13} = \dfrac{7}{-3b + 1}$

11) $\dfrac{-11}{100} = \dfrac{7.92}{y}$

12) $3x + 2 = -5x + 2$

13) $0.38x + 2.4 = 10.2 - 1.57x$

14) $53 = \dfrac{k - 13}{4}$

15) $9(x + 3) - 4(2x - 5) = 3(7x - 6)$

16) $\dfrac{3}{4}m + \dfrac{1}{4} = \dfrac{5}{4}$

17) The average of 28, 35, 39 and x is 37. Find the value of x.

18) The length and the width of a rectangle are 2x + 9 and x − 7, respectively. If the perimeter is 58, find the area of the rectangle.

19) To fix your air conditioner, the repairman charges $32 per hour. Also, you need to pay $150 for the parts. If you paid a total of $262, how many hours did the repair work take?

20) Jennifer is five years older than twice Peter's age. Their combined ages is 56. How old is Jennifer?

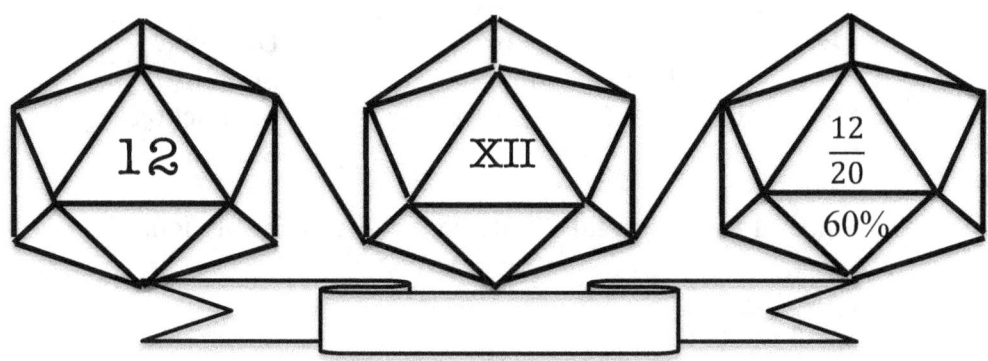

Lesson 12: More Equations and Formulas

In the previous lesson, you solved linear equations that all have one solution. However, it is possible that a linear equation has no solution, or more than one solution.

Example 1: Solve.

a) $x + 3 = x + 4$

b) $2(3x - 7) = 4 - (5 - 6x)$

c) $\dfrac{x}{10} = \dfrac{2x+1}{20}$

Solution: a) In this problem, we are looking for a number such that when we add 3 to it, it is equals to that same number plus 4. There is no such a number because the left hand side will always be one less than the right hand side.

We could also solve this problem as before. We have,
$$x + 3 = x + 4$$
$$\Rightarrow \quad x + 3 - x = x + 4 - x \quad \text{Subtract x on both sides}$$
$$\Rightarrow \quad 3 = 4 \quad \text{Simplify}$$

Since $3 = 4$ is nonsense, the result is no solution.

b) This problem is not as obvious as the previous one. So, we can proceed normally as before. We have,

61

$$2(3x - 7) = 4 - (5 - 6x)$$
=> $\quad 6x - 14 = 4 - 5 + 6x \quad$ Distribute
=> $\quad 6x - 14 = -1 + 6x \quad$ Simplify
=> $\quad 6x - 14 - 6x = -1 + 6x - 6x \quad$ Subtract 6x
=> $\quad -14 = -1 \quad$ Simplify

Since $-14 = -1$ is a false statement, there is no solution.

c) $\dfrac{x}{10} = \dfrac{2x+1}{20}$

$20x = 20x + 10 \quad$ Cross Multiply
$20x - 20x = 20x + 10 - 20x \quad$ Subtract 20x on both sides
$0 = 10 \quad$ Simplify

Again, since $0 = 10$ is NOT true, there is no solution.

Example 2: Solve.

a) $5x + 1 = 5x + 1$ 　　　b) $\dfrac{1}{4}m - 21 = 4 - \dfrac{1}{2}(50 - \dfrac{1}{2}m)$

Solution: 　a) In this problem, the left hand side looks exactly like the right hand side. This is called an **identity**. We can plug in any numbers for x, and the equation is true. For example, if $x = 1$, both sides equal $5(1) + 1 = 6$. If $x = 2$, both sides equal $5(2) + 1 = 11$. So, there are **infinitely many solutions**, or the solutions are **all real numbers**.

b) $\dfrac{1}{4}m - 21 = 4 - \dfrac{1}{2}(50 - \dfrac{1}{2}m)$

$\dfrac{1}{4}m - 21 = 4 - 25 + \dfrac{1}{4}m \quad$ Distribute

$\dfrac{1}{4}m - 21 = -21 + \dfrac{1}{4}m \quad$ Simplify

$\dfrac{1}{4}m - 21 - \dfrac{1}{4}m = -21 + \dfrac{1}{4}m - \dfrac{1}{4}m \quad$ Subtract $\dfrac{1}{4}m$ on both sides

$-21 = -21 \quad$ Simplify

Since $-21 = -21$ is a true statement, there are infinitely many solutions.

A formula such as distance = rate × time, or $d = rt$ is an equation that has more than one variable. In this case, we can solve for one of its variables in terms of the others.

Example 3: Solve each formula for the indicated variable.

    a) $d = rt$, for r b) $P = 2l + 2w$, for w

    c) $A = \frac{1}{2}bh$, for h d) $F = \frac{9}{5}C + 32$, for C

Solution:

a) To solve for r, we have to isolate r.

$$d = rt$$
$$\frac{d}{t} = \frac{rt}{t} \quad \text{Divide by t on both sides}$$
$$\boxed{\frac{d}{t} = r} \quad \text{Simplify}$$

b) $P = 2l + 2w$
$$P - 2l = 2l + 2w - 2l \quad \text{Subtract 2l on both sides}$$
$$P - 2l = 2w \quad \text{Simplify}$$
$$\frac{P-2l}{2} = \frac{2w}{2} \quad \text{Divide by 2 on both sides}$$
$$\boxed{\frac{P-2l}{2} = w} \quad \text{Simplify}$$

c) $A = \frac{1}{2}bh$
$$2A = 2(\tfrac{1}{2}bh) \quad \text{Multiply by 2 on both sides}$$
$$2A = bh \quad \text{Simplify}$$
$$\boxed{\frac{2A}{b} = h} \quad \text{Divide by b on both sides}$$

d) $F = \frac{9}{5}C + 32$
$$F - 32 = \tfrac{9}{5}C + 32 - 32 \quad \text{Subtract 32 on both sides}$$
$$F - 32 = \tfrac{9}{5}C \quad \text{Simplify}$$
$$\tfrac{5}{9}(F - 32) = \tfrac{5}{9}(\tfrac{9}{5}C) \quad \text{Multiply by } \tfrac{5}{9} \text{ on both sides}$$
$$\boxed{\tfrac{5}{9}(F - 32) = C} \quad \text{Simplify}$$

Practice 12

Solve.

1) $-a - 32 = -32 - a$

2) $1.5(2x + 6) = 18 + 3x$

3) $2(b - 5) = b - 10$

4) $\frac{1}{4} + 2c = 2(c + \frac{1}{8})$

5) $\frac{2m-1}{8} = \frac{m}{4}$

6) $\frac{x}{3} - 4 = \frac{5x}{6} - 4$

7) $2.8y - 2(1.4y + 2) = 2$

8) $-(1 - 4k) = 2(2k + 3) - 7$

9) $\frac{2}{-0.2n - 3.2} = \frac{-10}{16 + n}$

10) $9 = \frac{-9p}{p - 2}$

Solve each formula for the indicated variable.

11) $I = Prt$, for t

12) $y = mx + b$, for x

13) $x + xy = 10$, for y

14) $E = mc^2$, for m

15) $4x - 6y = 8$, for y

16) $\frac{3}{4}xyz + \frac{1}{4} = xy$, for z

17) $A = \frac{1}{2}(b_1 + b_2)h$, for b_1

18) $V = lwh$, for w

19) $y - y_1 = m(x - x_1)$, for x

20) $ax + by = c$, for y

Bonus: Solve for x. $ax + by = cz + dx$

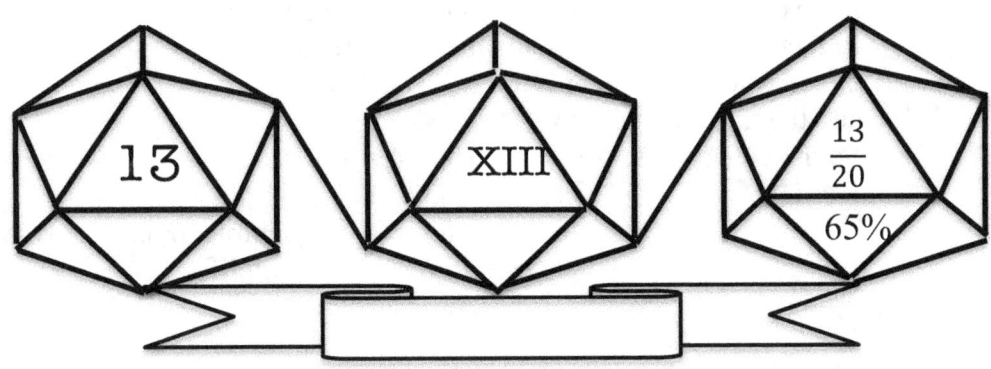

Lesson 13: Word Problems

To study math is to study problem solving. Therefore, after you learned how to solve equations, it is very important that you are able to apply them to word problems. You have already encountered a few word problems in some of the earlier lessons. In this lesson, we will continue to examine a variety of word problems that are commonly seen in algebra.

Example 1: The sum of three **consecutive** integers is 72. What are the integers?

Solution: "Consecutive" means "one after another".
For example, 3, 4, and 5 are consecutive integers.
In fact, if we let x represents the first integer,
then the second integer is x + 1,
and the third integer is x + 2.
So the equation is: x + x + 1 + x + 2 = 72.

$3x + 3 = 72$ Simplify
$3x + 3 - 3 = 72 - 3$ Subtract 3 on both sides
$3x = 69$ Simplify
$x = 23$ Divide by 3

Hence, the three integers are: 23, 24, and 25.

Example 2: The sum of three **consecutive odd** integers is 183. What is the largest integer?

Solution: "Consecutive odd" means "one odd number after another".

For example, 3, 5, and 7 are consecutive odd integers.
In fact, if we let x represents the first integer,
then the second integer is x + 2,
and the third integer is x + 4.
So the equation is: x + x + 2 + x + 4 = 183.

$$3x + 6 = 183 \quad \text{Simplify}$$
$$3x + 6 - 6 = 183 - 6 \quad \text{Subtract 6 on both sides}$$
$$3x = 177 \quad \text{Simplify}$$
$$x = 59 \quad \text{Divide by 3}$$

Now since we let x represents the first or smallest integer, the largest integer is x + 4 = 59 + 4 = **63**.

This brings up an important point. When you solve a word problem, make sure you answer the question at the end. The value of x may or may not be the final answer.

In fact, on the SAT, a lot of multiple-choice questions make you solve for x, then ask for something else. So, you have to be careful of the traps.

If you want extra practice and to explore this problem more, here is another way to solve it.
Let x represents the largest integer.
Then x – 2 is the second integer,
and x – 4 is the smaller integer.
So the equation is: x + x – 2 + x – 4 = 183

$$3x - 6 = 183 \quad \text{Simplify}$$
$$3x - 6 + 6 = 183 + 6 \quad \text{Add 6 on both sides}$$
$$3x = 189 \quad \text{Simplify}$$
$$x = 63 \quad \text{Divide by 3}$$

This time x = 63 is the answer.

Example 3: At noon, a taxi leaves LAX traveling at a rate of 55 mi/h. At the same time, a second taxi leaves the same location traveling at 45 mi/h in the opposite direction. At what time will the taxis be 250 miles apart?

Solution: A lot of students find this type of problems challenging, even though they know they have to somehow use the formula d = rt. The reason is, it seems as though both the distances and times are missing; only the rates are given. Here is how you can solve it.

Let t be the amount of time it takes for the taxis to be 250 miles apart.

Since the first taxi is traveling at 55mi/h for t hours, the distance is **d = 55t**. Again, look like we are missing two variables.

However, we can set up another equation for the second taxi.

Since the second taxi is traveling at 45 mi/h for t hours, the distance is **d = 45t**.

Since the two taxis traveled in opposite direction, you add their distances and set it equals to 250 miles. Below is the diagram to make it more clear.

So the equation is 55t + 45t = 250.
Combine like terms, we have 100t = 250. Hence, **t = 2.5 hours**.
That means at **2:30 PM**, the taxis will be 250 miles apart.

Example 4: Suppose the two taxis in example 3 traveled in the same direction. How long will it take for the two taxis to be 30 miles apart?

Solution: Below is the new diagram for this situation.

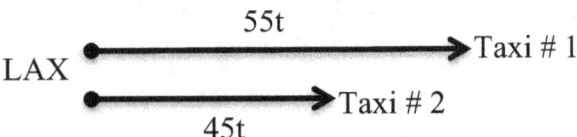

In this case, you will need to subtract their distances and set it equals to 40 miles. So the equation is 55t − 45t = 30. Combine like terms, we have 10t = 30. Hence, **t = 3 hours**.

To check your answer: Taxi # 1 traveled 55 × 3 = 165 miles. Taxi # 2 traveled 45 × 3 = 135 miles. 165 − 135 = 30 miles.

Example 5: There are three times as many pennies as dimes. There are 2 more dimes than quarters. If the total value of the coins is $3.68, how many penny are there?

Solution: Since the amount of pennies equals three times the amount of dimes, you can think of this as "**pennies depend on dimes**." In other words, we need to know the amount of dimes before we can figure out the pennies.

Similarly, since the number of dimes equals the number of quarters plus two, that means "**dimes depend on quarters**." We need to know the number of quarters first.

Let q be the number of quarters.
The number of dimes is q + 2.
The number of pennies is 3 × (q + 2).

Now, since each quarter is 25 cents, q quarters equals q × 25 cents. Similarly, (q + 2) dimes equals (q + 2) × 10 cents.
So the equation is 25q + 10(q + 2) + 3(q + 2) = 368 cents.

=>	25q + 10q + 20 + 3q + 6 = 368	Distribute
=>	38q + 26 = 368	Simplify
=>	38q + 26 − 26 = 368 − 26	Subtract 26
=>	38q = 342	Simplify
=>	**q = 342 ÷ 38 = 9**	Divide 38

So, there are 9 quarters. That means there are 11 dimes.
Hence, the number pennies is **33**.

Example 6: The measures of the angles of a triangle are in the ratio 2 to 3 to 4. Use the fact that the sum of the measures of the three angles in a triangle is always 180 degree, find measure of the largest angle.

Solution: If a and b are in the ratio 2 to 3, that could mean a = 2 and b = 3. But, it could also mean a = 4 and b = 6, or a = 6 and b = 9, or etc... In general, a must equal 2x, and b must equal 3x.

Hence, the three angles are: 2x, 3x, and 4x. So, the equation is 2x + 3x + 4x = 180. That means we have 9x = 180. Therefore, x = 20. So, the largest angle is 4x = 4(20) = **80 degree**.

Practice 13

1) The sum of four consecutive integers is 202. What are the integers?

2) The sum of three consecutive even integers is 84. What is the middle integer?

3) There are twice as many nickels as quarters. There are 5 fewer quarters than dimes. If the total value of the coins is $3.20, how many quarters are there?

4) Ted is 11 years younger than James. Michelle is three times as old as Ted. The total ages of the three people is 46 years old. How old is James?

5) Three sides of a triangle are in the ratio 4 to 5 to 8. If the perimeter of the triangle is 85. What is the measure of the longest side?

6) There are 16 girls in the class. This is eight less than twice the number of boys. If each student contributes $2.50 to have a pizza party, how much money will be available?

7) Tom travels from point A to point B at a rate of 57 mi/h. When he returns from point B to point A, he moves 19 mi/h faster, and as a result takes one hour less time. What is the distance from A to B?

8) Lin is three years older than twice Michael's age. Five years from now, their combined age will be 64. How old will Lin be next year?

9) At 7:00 A.M., a car leaves company XYZ traveling at a rate of 36 mi/h. Half an hour later, a second car leaves the same location traveling at a rate of 48mi/h in the same direction. At what time will the second car catch up to the first car?

10) In two hours, Melissa can drain 320 gallons of water out of a pool. At this rate, how long will it take to drain a 10000-gallon pool?

11) Yesterday, Anthony made seven of ten free throws. How many more free throws must Anthony make in a row, for his percentage to be 80%?

12) The length of a rectangle is 14cm less than 5 five times the width. If you increase the width by 6cm and the length by 2cm, the ratio of the width to length is 1 to 2. What is the perimeter of the original rectangle?

13) A piece of rope is 32 feet long. Danny cuts it into two pieces so that one piece is 9 feet longer than the other. How long is the shorter piece?

14) A school performance cost $10 for an adult and $6 for a child. There are twice as many children attend as adults, and the total amount of money collected from selling the tickets is $2244. How many children see the performance?

15) A salesman has a base salary of $1000 per month, plus a 5% commission on his sales. How much must he sell to have an income of $2500 per month?

16) Using his small car, Evan can drive 45 miles per gallon of gas. Using his SUV, Evan can only drive 20 miles per gallon. If a gallon of gas cost $3.00 and Evan drives 1800 miles using his small car, how much money will he save?

17) An online music company charges a membership fee of $5 per month, plus 35 cents for each song a customer downloads. If Janet has a $20 budget, what is the most number of songs she can download in a month?

18) A phone company charges 75 cents per minute for the first 15 minutes, and 55 cents per minute for each additional minute. If Max has $30 to spend, what is the most number of minutes he can talk on the phone?

19) One integer is 3 more than twice the other. The average of the two integers is 30. What is the difference between those two integers?

20) Christy cuts a rope into 3 pieces. The first piece is 8m longer than the second piece. The third piece is twice as long as the first piece. If the average of the three pieces is 16m, how long is the third piece?

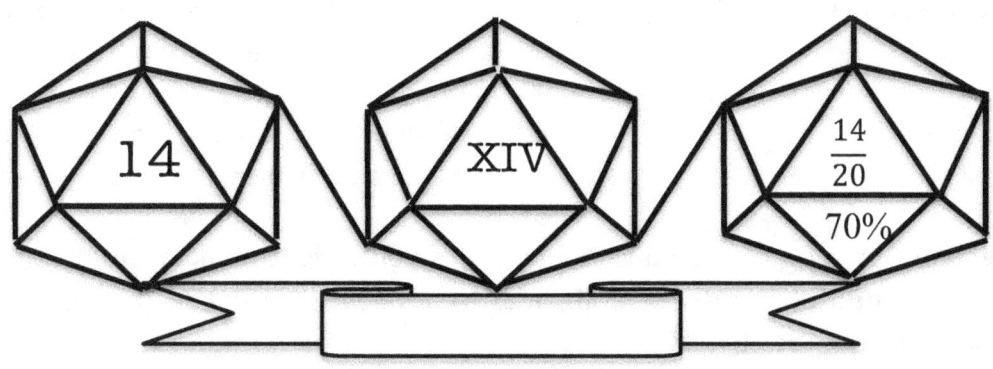

Lesson 14: Solving Linear Inequalities

In lesson 11, you learned that setting 2 algebraic expressions equal to each other produces an equation. In this lesson, we will set one expression < (less than), > (greater than), ≤ (less than or equal), or ≥ (greater than or equal) to another. The result is, you have an **inequality**.

Solving a linear inequality, or finding values of the variable that make the inequality true, is similar to solving a linear equation. As before, you can use properties to help you move from one inequality to another simpler and equivalent inequality.

The Addition Property of Inequality: If a < b, then a + c < b + c, and
If a > b, then a + c > b + c.

The Subtraction Property of Inequality: If a < b, then a − c < b − c and
If a > b, then a − c > b − c.

The Multiplication Property of Inequality: If a < b and c > 0, then ac < bc, and
If a > b and c > 0, then ac > bc.

If a < b and c < 0, then ac > bc, and
If a > b and c < 0, then ac < bc.

The Division Property of Inequality: If $a < b$ and $c > 0$, then $a \div c < b \div c$, and
If $a > b$ and $c > 0$, then $a \div c > b \div c$.

If $a < b$ and $c < 0$, then $a \div c > b \div c$, and
If $a > b$ and $c < 0$, then $a \div c < b \div c$.

Important: As you can see, the "extra" rules you have to remember here are: "When you **multiply or divide** both sides **by a negative number**, you have to **switch the inequality sign**."

Example 1: Solve each inequality and graph the solution.

    a) $x + 4 > 7$ b) $x - 3 < 1$

    c) $2x - 5 \leq -1$ d) $3(x + 4) \geq 21$

Solution: a) $x + 4 > 7$

    $x + 4 - 4 > 7 - 4$ Subtract 4 on both sides

    $\boxed{x > 3}$ Simplify

To graph the solution, you draw a number line and "shade" all of the numbers greater than 3. Since 3 is NOT greater than 3, 3 is NOT part of the solutions. You can draw an "open circle" to exclude 3.

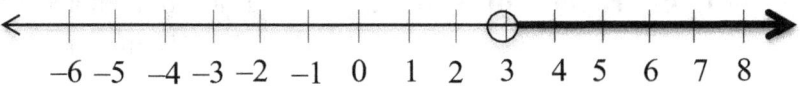

b) $x - 3 < 1$

    $x - 3 + 3 < 1 + 3$ Add 3 on both sides

    $\boxed{x < 4}$ Simplify

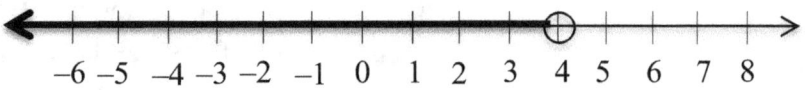

c) $2x - 5 \leq -1$

$2x - 5 + 5 \leq -1 + 5$ Add 5 on both sides

$2x \leq 4$ Simplify

$\boxed{x \leq 2}$ Divide by 2 on both sides

Since x is **less than or equal** to 2, 2 is a part of the solutions. To show that, you have to use a "closed" circle this time.

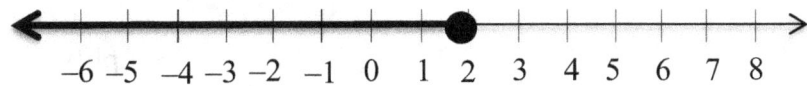

d) $3(x + 4) \geq 21$

$3x + 12 \geq 21$ Distribute

$3x + 12 - 12 \geq 21 - 12$ Subtract 12 on both sides

$3x \geq 9$ Simplify

$\boxed{x \geq 3}$ Divide by 3 on both sides

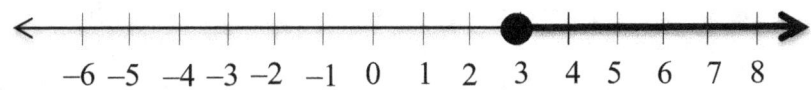

Example 2: Solve each inequality and graph the solution.

a) $-5x + 14 > 19$ b) $2 - x < 0$

c) $2x - 1 \leq 3x + 1$ d) $-7 \leq \frac{-2}{3}x - 5$

Solution: a) $-5x + 14 > 19$

$-5x + 14 - 14 > 19 - 14$ Subtract 14 on both sides

$-5x > 5$ Simplify

$\boxed{x < -1}$ Divide by negative 5

Notice that we have to switch the inequality sign from greater than to less than, when we divide both sides by a negative number.

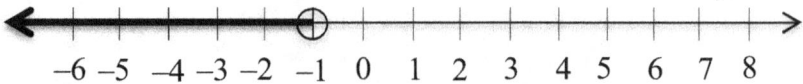

b) $2 - x < 0$

  $2 - 2 - x < 0 - 2$ Subtract 2 on both sides

  $-x < -2$ Simplify

  $\boxed{x > 2}$ Divide by negative 1

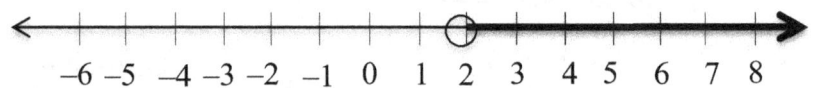

c) $2x - 1 \leq 3x + 1$

  $2x - 3x - 1 \leq 3x - 3x + 1$ Subtract 3x on both sides

  $-x - 1 \leq 1$ Simplify

  $-x - 1 + 1 \leq 1 + 1$ Add 1 on both sides

  $-x \leq 2$ Simplify

  $\boxed{x \geq -2}$ Divide by negative 1

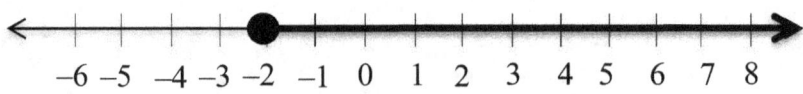

d) $-7 \leq \frac{-2}{3}x - 5$

  $-7 + 5 \leq \frac{-2}{3}x - 5 + 5$ Add 5 on both sides

$-2 \leq \dfrac{-2}{3}x$ \qquad Simplify

$3 \geq x$ \qquad Multiply by negative $\dfrac{3}{2}$

$\boxed{x \leq 3}$ \qquad write equivalent inequality

Note that the statement "3 is greater than or equal to x" is the same as "x is less than or equal to 3." (I.e., "You are younger than me." is the same as "I am older than you"). However, putting x on the left hand side makes it easier to think about the graph.

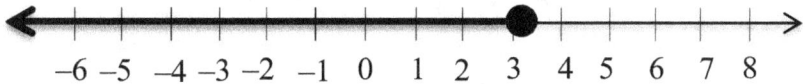

Example 3: Solve each inequality and graph the solution.

a) $x + 9 < x + 6$ \qquad b) $2x + 5 > 2(x - 1)$ \qquad c) $3x \leq 2x$

Solution: a) We are looking for a number such that when we add 9 to it, it is less than the same number add 6. There is no such a number. The left hand side is always 3 more than the right hand side. Since there is no solution, you leave the number line empty.

To show your work, you can also do the following:

$x + 9 < x + 6$

$x - x + 9 < x - x + 6$ \qquad Subtract x on both sides

$\boxed{9 < 6}$ \qquad Simplify

Since $9 < 6$ is false, there is no solution (and no graph).

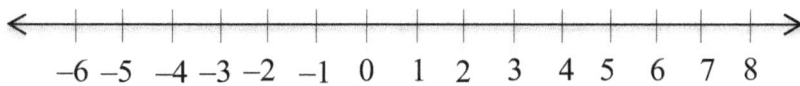

b) $2x + 5 > 2(x - 1)$

$\quad 2x + 5 > 2x - 2$ \hfill Distribute

$\quad 2x - 2x + 5 > 2x - 2x - 2$ \hfill Subtract 2x on both sides

$\quad \boxed{5 > -2}$ \hfill Simplify

Since $5 > -2$ is true, the solutions are **all real numbers**. In this case, you have to **"shade the entire number line."**

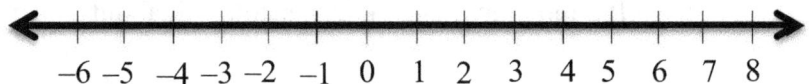

c) $3x \leq 2x$

$\quad 3x - 2x \leq 2x - 2x$ \hfill Subtract 2x on both sides

$\quad \boxed{x \leq 0}$ \hfill Simplify

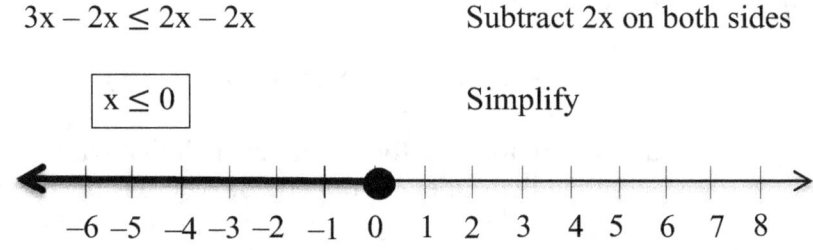

Practice 14

Solve each inequality and graph the solution.

1) $x + 11 > 8$

2) $25x - 56 \leq -106$

3) $-(x + 2) \geq -3$

4) $6x < -1.5 + 7x$

5) $\frac{3}{11}x - 4 < -4$

6) $x - 12 < -10 + x$

7) $1 - (6x - 4) \leq -2(3x + 2.5)$

8) $\frac{-1}{10}x + 1 \geq 1.3$

9) $-9 > -5 + x$

10) $15x - 2 \leq 14x - 2$

11) $\frac{x}{4} < \frac{x}{3}$

12) $\frac{x - 4}{5} > -0.3$

13) $-x > -x - 0.1$

14) $-0.8x \geq x - 0.9$

15) $\frac{-1}{4}x + \frac{1}{8} < \frac{5}{8}$

16) $2(3 - x) - 2(3x - 1) > 4(x + 8)$

17) $0 \geq 2(3 - x) + 2x - 6$

18) $10 > 5(x + 1) - 5x + 5$

19) The length of a rectangle is 5 more than twice the width. If the perimeter is greater than 40 m, what is the smallest possible integer value for the width?

20) Find all possible sets of three consecutive odd **whole** numbers whose sum is less than or equals to 24.

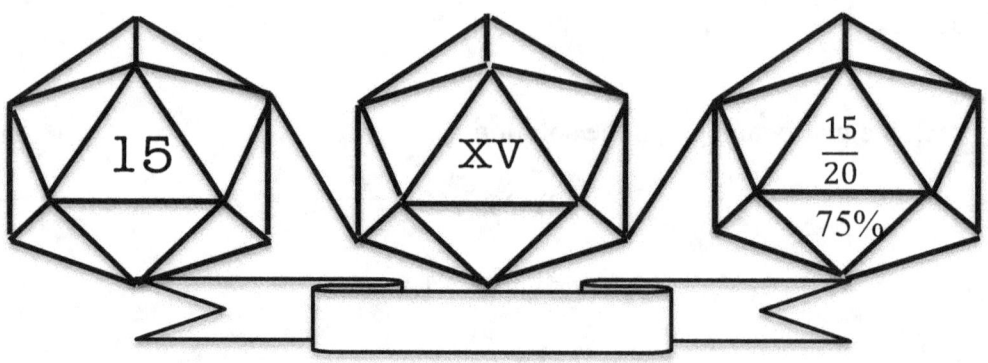

Lesson 15: Introduction To Graphs

In the previous lesson, we used a number line to graph the solution set of a linear inequality in one variable. In this lesson, we will make two number lines intersect at right angles in order to form a **coordinate plane**. We will call the horizontal number line, the **x-axis**, and the vertical number line, the **y-axis**. The two lines will intersect at a point called the **origin**. Here is an example of what a coordinate plane looks like.

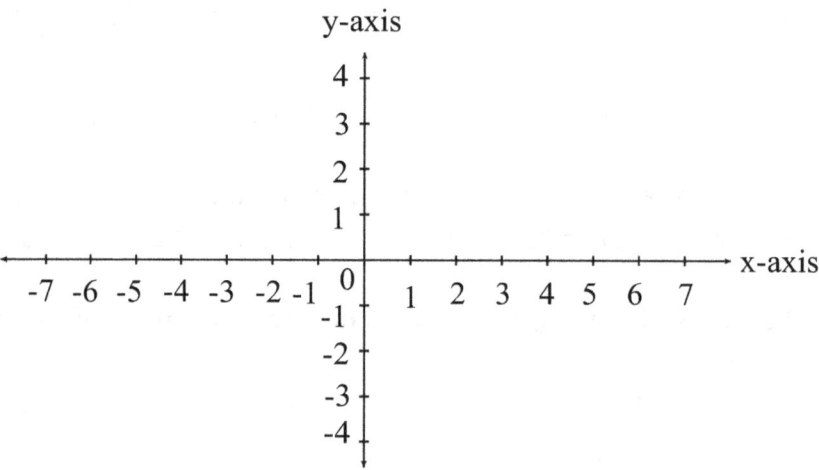

To locate a point in the plane, we use an **ordered pair** (x, y). For example, the point (5, 2) is located at 5 units to the right and 2 units up from the origin. It is an ordered pair because the order matters. For example, the point (2, 5) is 2 units to the right and 5 units up, which is different from the point (5, 2).

Pre-Algebra

Example 1: Plot and label the following points in the coordinate plane:

L(–1, –3), M(0, 4), N(1, –3), P(–2, 4), Q(3, 0).

Solution: For point L, go **left** 1 unit and **down** 3 units from the origin. For point M, go **up** 4 units. There is no left or right movement because the x-coordinate is 0.

Here are the rest of the points on the graph.

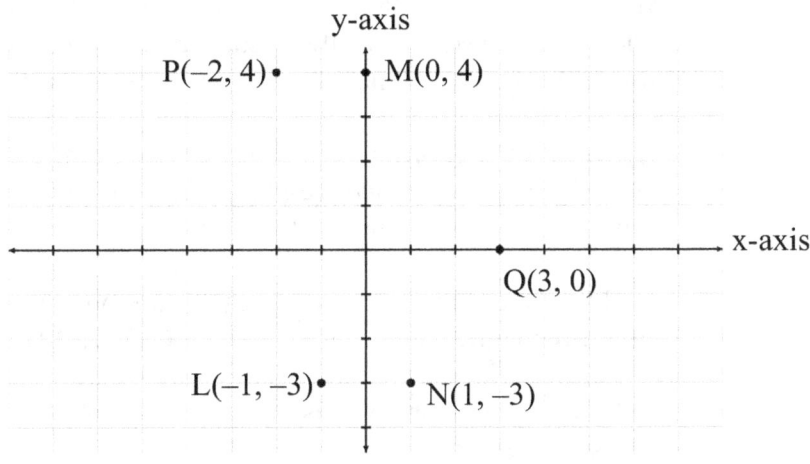

In a coordinate plane, there are **four quadrants**. The first quadrant starts at the upper right hand region; the second quadrant is at the upper left. As you continue to move in the counterclockwise direction, the third quadrant is located at the lower left, and the fourth is at the lower right. Here is the picture.

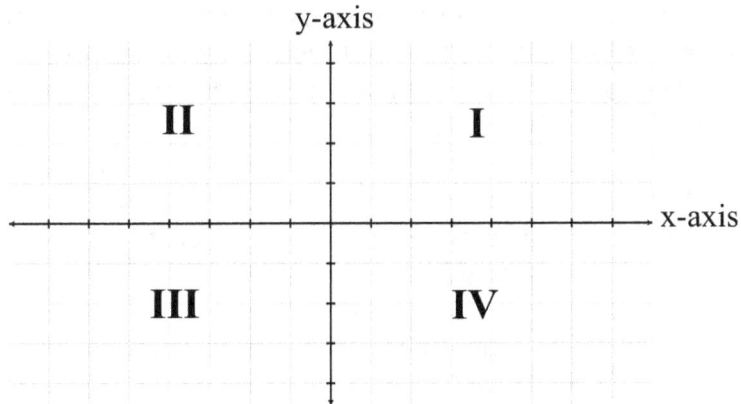

Important: The signs of the coordinates tell you which quadrant a point is in.

79

Example 2: State the quadrant where each point is located.

    a) (2, −7)   b) (−10, 5)   c) (−6, −4)

    d) (1.5, 2.6)   e) (0, 1)   f) (2, 0)

Solution: a) Since the x-coordinate is positive, the point is on the **right** hand side of the y-axis. Since the y-coordinate is negative, the point is **below** the x-axis. Hence, (2, −7) is in quadrant IV.

b) Since the x-coordinate is negative, the point is on the **left** hand side of the y-axis. Since the y-coordinate is positive, the point is **above** the x-axis. Hence, (−10, 5) is in quadrant II.

c) Since the x-coordinate is negative, the point is on the **left** hand side of the y-axis. Since the y-coordinate is negative, the point is **below** the x-axis. Hence, (−6, −4) is in quadrant III.

d) Since both the x and y-coordinates are positive, the point is on the **upper right** hand region. Hence, (1.5, 2.6) is in quadrant I.

e) Since the x-coordinate is 0, this point is NOT in any quadrant. It lies on the y-axis.

f) Since the y-coordinate is 0, this point is NOT in any quadrant. It lies on the x-axis.

The solution to a one-variable equation such as $x + 5 = 9$ is a number. In this case, the solution is $x = 4$. For an equation with two variables such as $y = x + 3$, the solutions consist of infinitely many pairs of numbers. For example, one solution is $x = 1$, and $y = 4$. Another solution is $x = 2$, and $y = 5$. Still another solution is $x = 0$, and $y = 3$. Notice that we can describe these solutions using the ordered pairs (1, 4), (2, 5), and (0, 3).

Important: Since the solutions of a two-variable equation are ordered pairs, not only we can put them on a graph, we can also connect the points to get a better visualization of the overall trend of our solutions.

Example 3: Find five solutions and graph $y = 2x + 1$.

Solution: Start by picking a number for x and substitute it into the equation to solve for y. In fact, x is the **independent variable** (i.e., you can pick whatever you want for x), and y is the **dependent variable** (i.e., what you get for y depends on what you picked for x).

To make it easier, suppose x = 0, then y = 2(0) + 1 = 1. So, **(0, 1)** is a solution. Next, if x = 1, then y = 2(1) + 1 = 3. So, **(1, 3)** is a second solution. Continue picking x = 2, we have y = 2(2) + 1 = 5. Hence, **(2, 5)** is a third solution.

Since you should also try to pick some negative numbers, suppose x = –1, then y = 2(–1) + 1 = –2 + 1 = –1. So, (–1, –1) is a fourth solution. Finally, let x = –2, then y = 2(–2) + 1 = –4 + 1 = –3. Therefore, (–2, –3) is our fifth solution.

Here is a summary of the five solutions in a table.

x	–2	–1	0	1	2
y	–3	–1	1	3	5

Here is the graph of the equation consisting of these points.

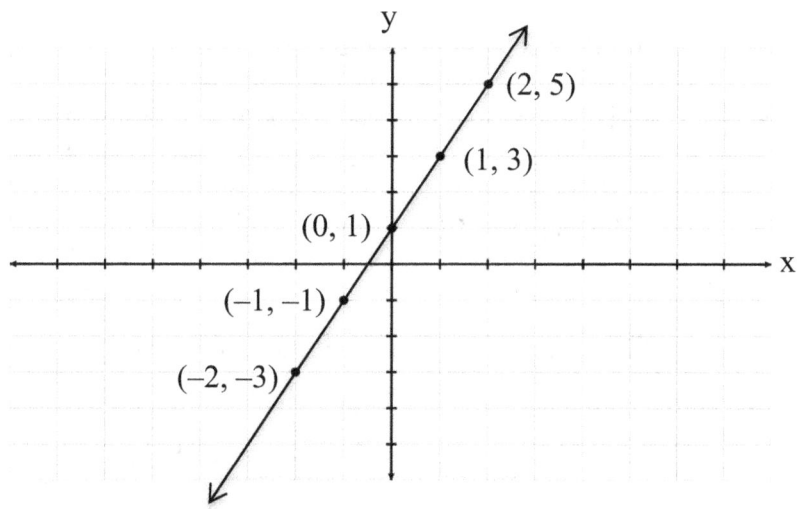

As you can see, in addition to our five solutions, all of the other points on this line give you all of the possible solutions of this equation, and that is the trend.

Example 4: Find three solutions and graph y = –3x – 1.

Solution: For x = 0, y = –3(0) – 1 = 0 – 1 = –1.
For x = 1, y = –3(1) – 1 = –3 – 1 = –4.
For x = –1, y = –3(–1) – 1 = 3 – 1 = 2.

81

Here is the table of values.

x	−1	0	1
y	2	−1	−4

Here is the graph of the equation consisting of these points.

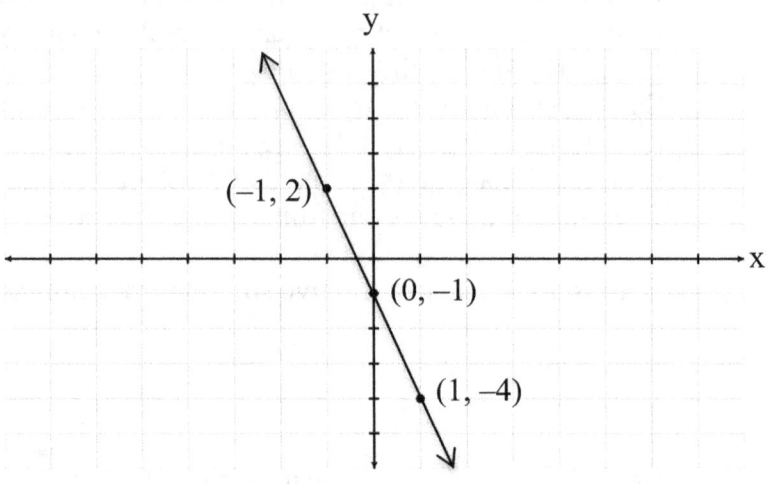

Example 5: Find three solutions and graph y = 4.

Solution: In this problem, y is a constant. It is equal 4 for all values of x. So, when x = 0, y = 4. When x = 2, y = 4. When x = −2, y = 4.

Here is the table of values.

x	−2	0	2
y	4	4	4

Here is the graph of the equation consisting of these points.

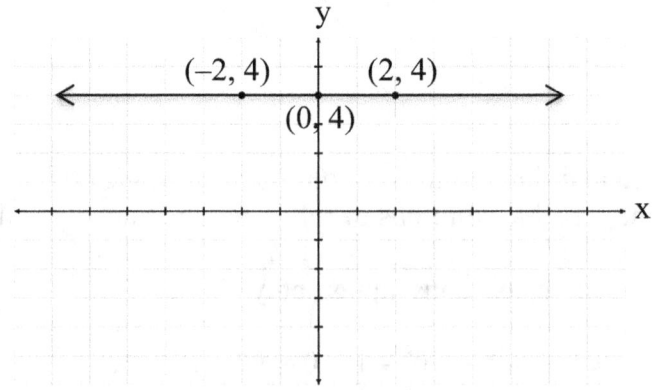

Note that when y is a constant, the graph is always a horizontal line.

Example 6: Find three solutions and graph x = 5.

Solution: In this problem, x is a constant. It is always equal 5. This is one of the rare instances where you CANNOT randomly choose a value for x. You have to pick numbers for y instead. Now, when y = 0, x = 5. When y = 2, x = 5. When y = –2, x = 5.

Here is the table of values.

x	5	5	5
y	–2	0	2

Here is the graph of the equation consisting of these points.

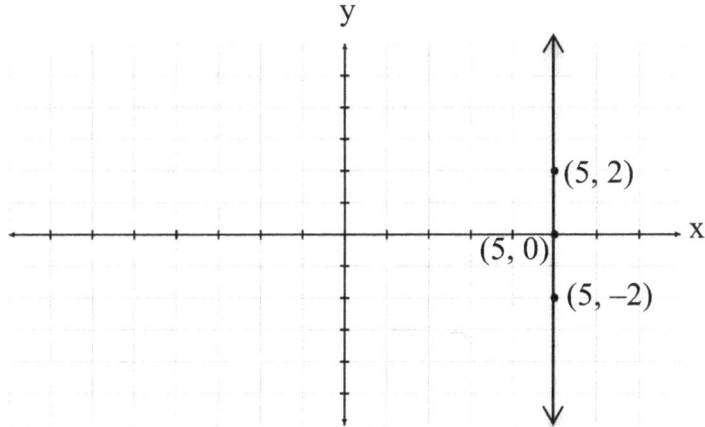

Note that when x is a constant, the graph is always a vertical line.

You will learn a lot more about graphing different forms of linear equations in Algebra I. Hopefully, this lesson will serve as a good introduction to graphing for you.

Practice 15

Plot and label each point in the coordinate plane.

1) (0, –3) 2) (4.5, 0)

3) (–3, –4) 4) (–2, 1)

State the quadrant where each point is located.

5) (9, –8) 6) (–10, 0)

7) (–2.3, –4.5) 8) (–0.3, 0.1)

9) (0, 0) 10) $(\frac{1}{8}, \frac{1}{3})$

Find three solutions and graph each equation.

11) $y = x$ 12) $y = -x$

13) $y = 2x - 1$ 14) $y = \frac{1}{2}x + 1$

15) $y = \frac{-1}{2}x + 2$ 16) $y = -2x$

17) $y = -2$ 18) $x = 1$

19) $y + x = 2$ 20) $x - y = 1$

Bonus: Graph $2x + 3y = 6$.

Pre-Algebra

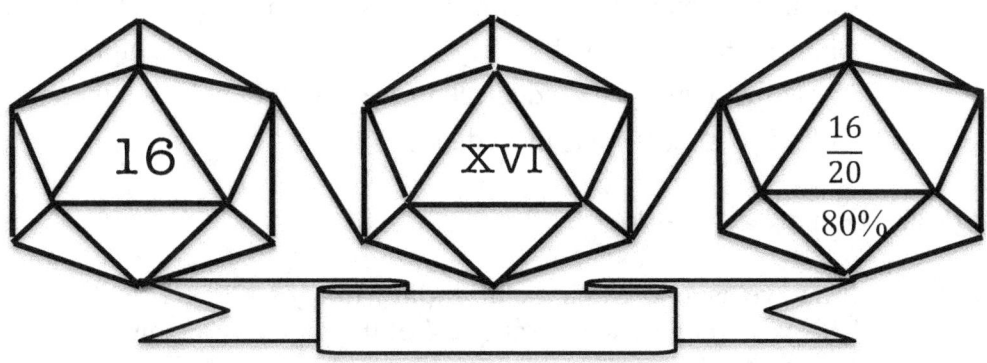

Lesson 16: Graphing Linear Inequalities in Two Variables

The graph of a linear equation in two variables is a straight line that contains all of the solutions of the equation. The regions above and below the line are **half-planes** that are each represented by the graph of a linear inequality.

Example 1: Graph $y \leq 2x + 1$ on a coordinate plane.

Solution: First, we have to graph the **boundary line** $y = 2x + 1$.
This is already done in example 3 of the previous lesson.
Next, we have to shade the half-plane **below** the line because we want y to be **less than or equal** to $2x + 1$. Here is the graph.

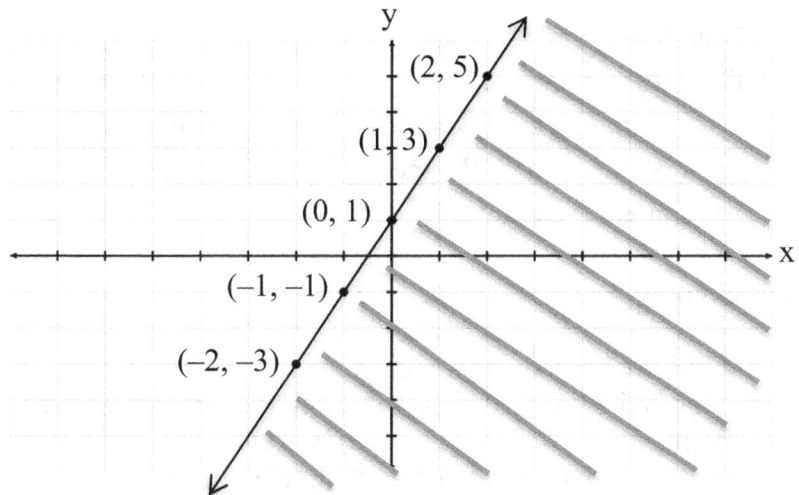

20/20 Math

Note that if you pick any points in the shaded region, (including the points on the line), it will satisfy the inequality. For example, let's choose the point (4, 3) and check to see if $y \leq 2x + 1$. In other words, is $3 \leq 2(4) + 1$? Indeed, $3 \leq 9$ is true.

Example 2: Graph $y > -3x - 1$.

Solution: First, we have to graph the **boundary line** $y = -3x - 1$. This is already done in example 4 of the previous lesson. However, this time we have to draw a broken or **dash line** to show that points on this line are not parts of the solutions. Then, we have to shade the half-plane **above** the line because we want y to be **greater than** $-3x - 1$. Here is the graph.

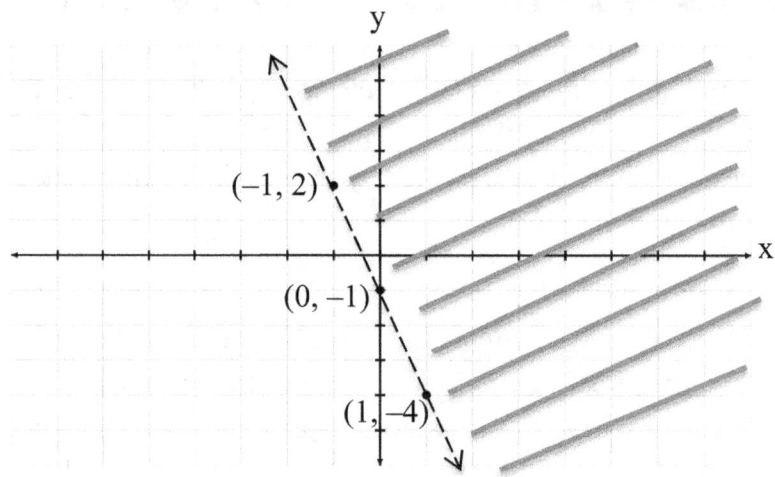

Note that if you are not sure which region to shade after you draw the line, you can pick a point and plug it into the inequality. If it satisfies the inequality, then shade the region containing the point. If it does not satisfy the inequality, then shade the region that does not contain the point.

Example 3: Graph $y > x$.

Solution: First, we have to draw a dash line for $y = x$. This is the line that passes through the origin and contains the points $(-2, -2)$, $(-1, -1)$, $(0, 0)$, $(1, 1)$ and $(2, 2)$. Next, we have to shade the half-plane **above** the line because we want y to be **greater than** x.

Here is the graph.

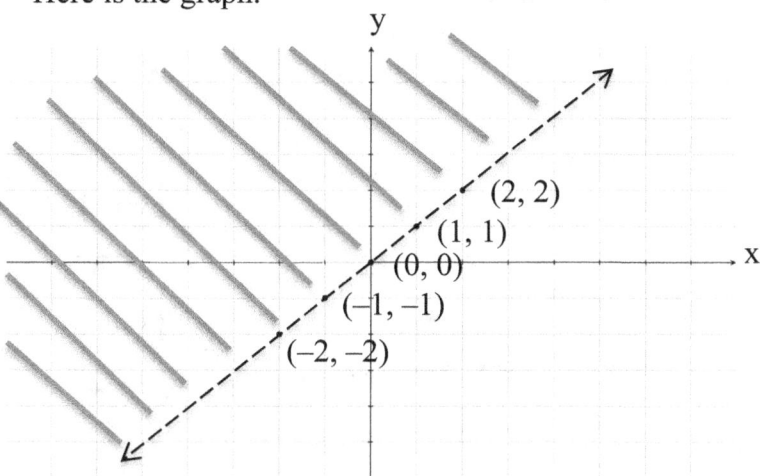

Example 4: Graph y ≤ 4.

Solution: First, we have to graph the **boundary line** y = 4. This is already done in example 5 of the previous lesson. Next, we have to shade the half-plane **below** the line because we want y to be **less than or equal** 4. Here is the graph.

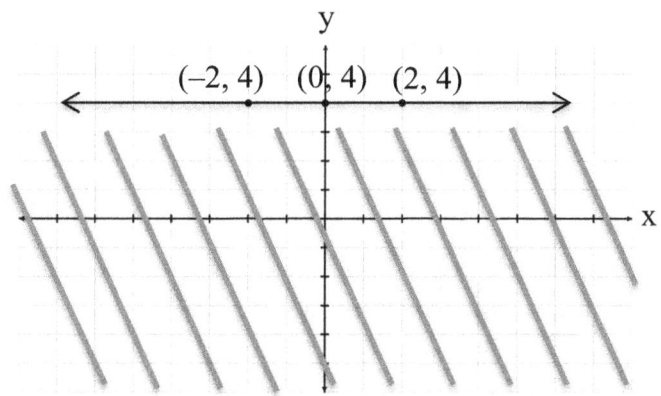

Example 5: Graph x < 5.

Solution: First, we have to graph the **boundary line** x = 5. This is already done in example 6 of the previous lesson, except we have to make it a dash line. Then, we have to shade the half-plane on the **left** of the line because we want x **less than** 5.

Here is the graph.

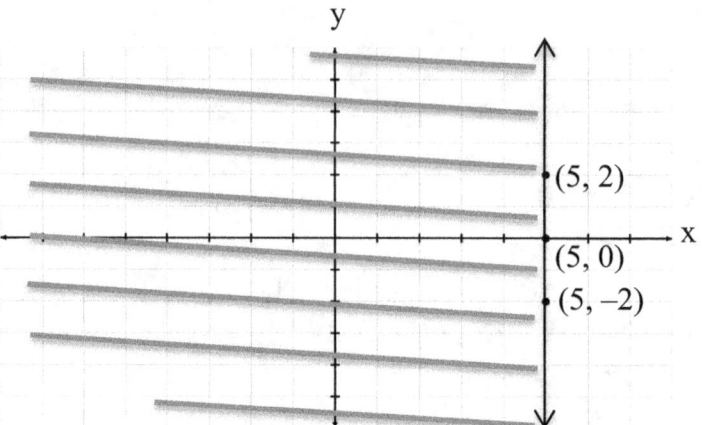

So as you can see, graphing linear inequalities in two variables is very similar to graphing linear equations. You just have to do an additional step of shading the correct region. In the next lesson, we will graph two lines on the same coordinate plane and find the intersection. In the meantime, have fun practicing.

Practice 16

Graph each inequality on a coordinate plane.

1) $y \leq x + 1$

2) $y \geq \frac{1}{2}x - 2$

3) $y < 3x$

4) $y > -2x$

5) $y \geq -x + 4$

6) $y \leq 0$

7) $x > 1.5$

8) $y < 2x - 3$

9) $y < -x$

10) $y \geq \frac{-1}{3}x + 1$

11) $0.5x < y$

12) $x < -2$

13) $y \geq 1 - 2x$

14) $y \leq \frac{2}{3}x$

15) $x + y < 2$

16) $y > -3x$

17) $y \leq -2 + x$

18) $1 - x > y$

19) $y > 1.5x + 1$

20) $x - y \geq 3$

Bonus: $4x - 3y < 12$.

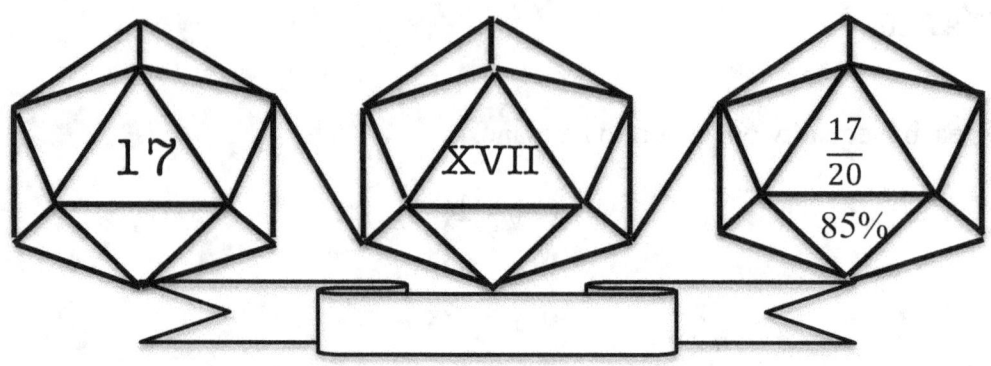

Lesson 17: Solving Systems of Equations by Graphing

Can you find two numbers such that the difference is 7 and the sum is 11? After thinking about it for a few minutes, you may be able to say that the first number is 9 and the second number is 2. However, besides using mental math, is there another way to approach this question? Well, solving this problem is equivalent to finding an ordered pair (x, y) that simultaneously satisfies the two equations $x - y = 7$ and $x + y = 11$. Graphically, we want to find a point (x, y) that lies on both lines. In other words, the intersection of the lines is the solution to the **system of equations** that we are trying to solve.

Example 1: Solve the system of equations by graphing:
$$y = 2x - 4$$
$$y = x - 1$$

Solution: Start by picking numbers for x and plug them into $y = 2x - 4$.

Here is the table of values.

x	−1	0	1
y	−6	−4	−2

Next, repeat the steps for $y = x - 1$.

Here is the table of values.

x	−1	0	1
y	−2	−1	0

Here is the graph of the two lines on the same coordinate plane.

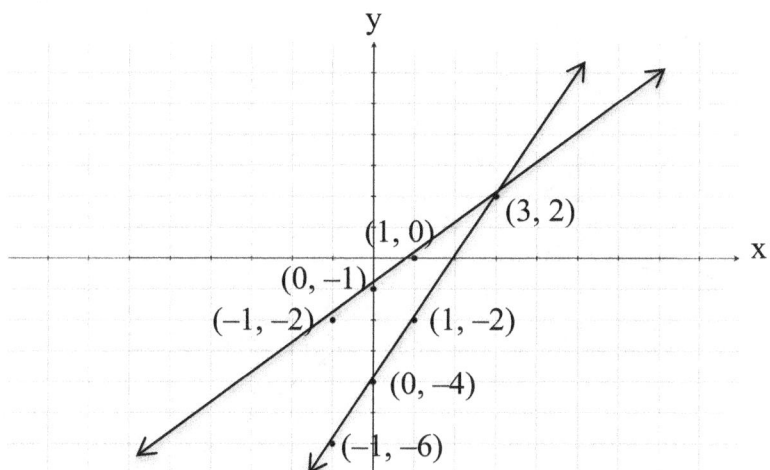

Since the lines intersect at (3, 2), the solution is **x = 3** and **y = 2**.

Let's check to see if the solution makes both equations true.
For the first equation: **2 = 2(3) − 4 = 6 − 4 = 2** is True.
For the second equation: **2 = 3 − 1 = 2** is True.

***Example 2**:* Solve the system of equations by graphing:
$$y = \frac{1}{3}x + 2$$
$$y = -x - 2$$

***Solution**:* For $y = \frac{1}{3}x + 2$, pick numbers that are multiples of 3 for x to avoid decimal answers.

Here is the table of values.

x	−3	0	3
y	1	2	3

For $y = -x - 2$, picking any integers is fine.

Here is the table of values.

x	−1	0	1
y	−1	−2	−3

91

Here is the graph of the two lines on the same coordinate plane.

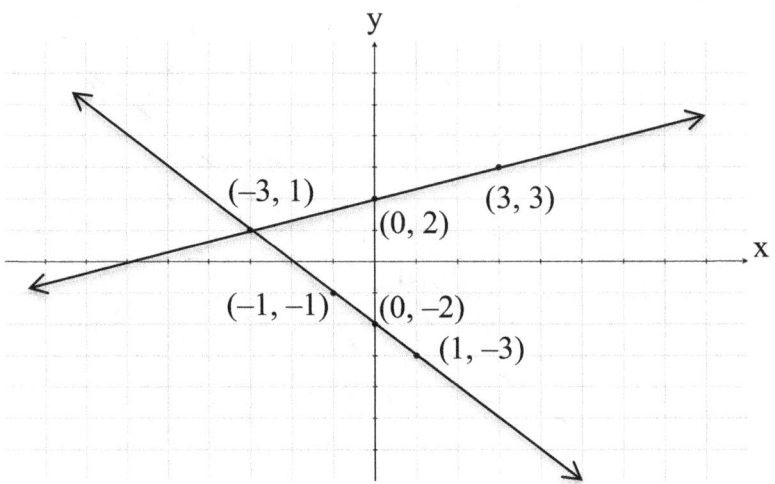

Since the lines intersect at (−3, 1), the solution is **x = −3** and **y = 1**.

Check:

For the first equation: $1 = \frac{1}{3}(-3) + 2 = -1 + 2 = 1$ is True.

For the second equation: $1 = -(-3) - 2 = 3 - 2 = 1$ is True.

Example 3:   Solve the system of equations by graphing:
$$y = 3x + 2$$
$$y = 3x - 2$$

Solution: Notice that these two equations look very similar. In fact, we are looking for two numbers such that 3 times the first number plus 2 is equal to the second number, AND simultaneously, 3 times the first number minus 2 is equal to the second number. Is it possible?

Let's graph both lines and see what will happen.

Here is the table for line 1.

x	−1	0	1
y	−1	2	5

Here is the table for line 2.

x	−1	0	1
y	−5	−2	1

Pre-Algebra

Here is the graph of the two lines.

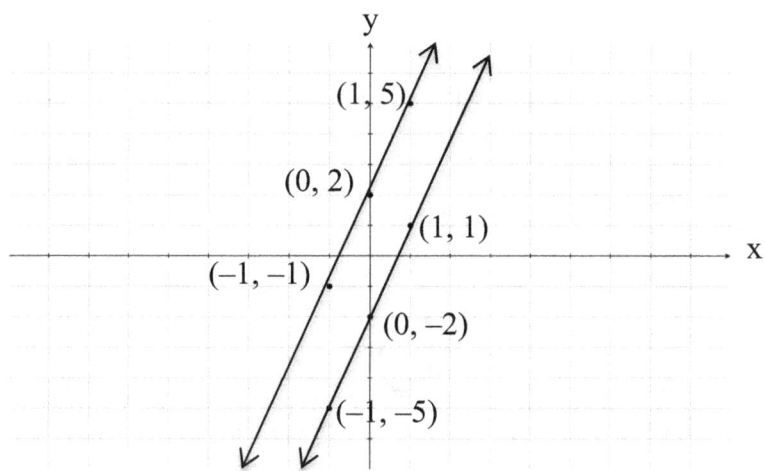

Since the lines are **parallel**, there is **no solution**. In fact, for every x-value, the first equation produces a y-value that is 4 more than the y-value of the second equation. For example, when x is 10, the first equation gives $y = 3(10) + 2 = 32$, and the second equation gives $y = 3(10) - 2 = 28$.

Important: A system of equations of the form $y = mx + b_1$ and $y = mx + b_2$, where $b_1 \neq b_2$, has no solution.

Example 4: Solve the system of equations by graphing:
$$4x + 2y = 2$$
$$y = -2x + 1$$

Solution: At first these two equations look very different. However, let's solve the first equation for y in terms of x so that at least the two equations are in the same form.

We have,	$4x + 2y = 2$	Given
=>	$-4x + 4x + 2y = -4x + 2$	Subtract 4x on both sides
=>	$2y = -4x + 2$	Simplify
=>	$y = -2x + 1$	Divide by 2 on both sides

As it turns out, the first equation is EXACTLY the same as the second equation. So, the two lines intersect infinitely many times. In other words, there is only one line, and there are infinitely many solutions.

Here is the table of values.

x	−1	0	1
y	3	1	−1

Here is the graph of the equations.

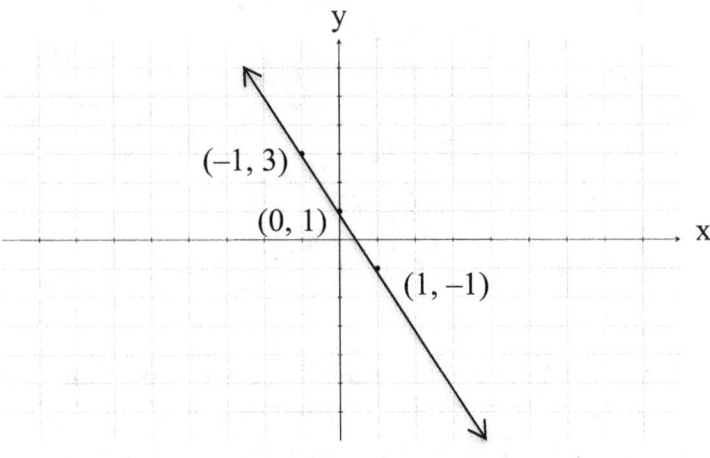

Example 5: Solve the system of equations by graphing:
$$y = 2$$
$$x = 4$$

Solution: This problem is quite simple. The first equation is a horizontal line, and the second is vertical. The intersection is (4, 2).

Here is the graph of the two lines.

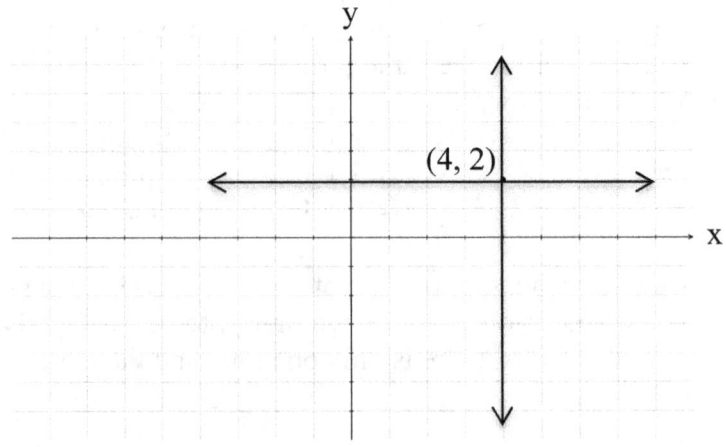

Practice 17

Solve each system of equations by graphing.

1) $y = 2x + 3$
   $y = -x + 6$

2) $y = \frac{1}{2}x - 1$
   $y = x - 2$

3) $y = 2$
   $y = x + 1$

4) $y = -2x - 2$
   $x = -3$

5) $y = -x + 2$
   $x + y = -4$

6) $y + 3 = 2x$
   $8x - 4y = 12$

7) $y = \frac{-2}{3}x + 4$
   $y = \frac{2}{3}x$

8) $y = 2 - x$
   $3x + 2y = 4$

9) $x + y = -1$
   $y = x + 1$

10) $y = 0$
    $y = -10$

11) $y = 1.5x - 2$
    $y = 0.5x$

12) $2x + 3y = 0$
    $-x + y = -5$

13) $y = 1 + 2x$
    $-x + y = 0$

14) $-x - 2y = 2$
    $2x + 4y = -4$

15) $12y = 24x + 36$
    $x = 0$

16) $0 = -4x + 2y$
    $y = x$

17) $-x = y$
    $5y = 1 - 5x$

18) $y = \frac{4}{5}x - 1$
    $3y + 6 = 3x$

19) $y = 2(x - 1.5)$
    $9y = 3(6x + 3)$

20) $3x - 4y = 2$
    $6x - 8y = 4$

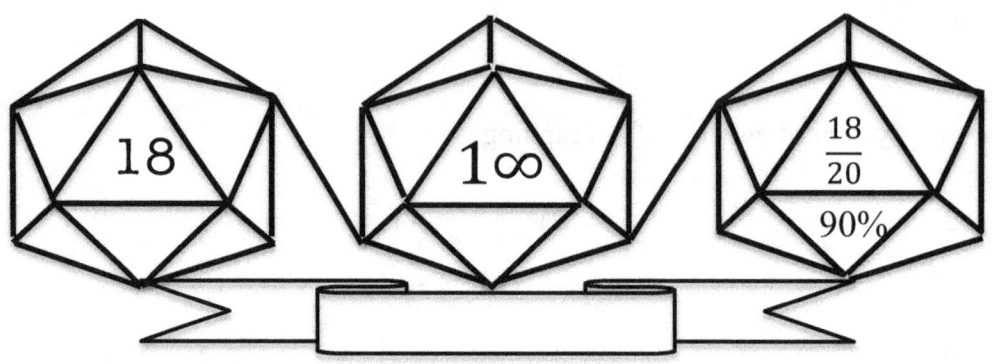

Lesson 18: Solving Systems of Equations by Substitution

Solving systems of equations by graphing only work when the solutions are integers. It would be nearly impossible to tell the intersection of the lines if it was a decimal or a fraction. Fortunately, there are algebraic ways to solve systems of equations. In this lesson, we will use the substitution method.

Example 1: Solve the system of equations by substitution:

$$8x + 3y = 9$$
$$y = 4x - 5$$

S_olution_: First look to see if there is an equation with a variable by itself. In this case, we have y **equals** 4x – 5. So, go into the first equation and substitute or replace y with 4x – 5. As a result, $8x + 3y = 9$ becomes $8x + 3(\mathbf{4x - 5}) = 9$. As we continue to solve this equation, we have

	$8x + 12x - 15 = 9$	Distribute
=>	$20x - 15 = 9$	Combine like terms
=>	$20x - 15 + 15 = 9 + 15$	Add 15 to both sides
=>	$20x = 24$	Simplify
=>	$x = \dfrac{24}{20} = \dfrac{6}{5} = 1.2$	Divide by 20

After getting x = 1.2, we can plug it back into either one of the orignal equations to solve for y. But since the second equation already has y by itself, it is easier to use that one. Thus, we have

$y = 4(1.2) - 5 = 4.8 - 5 = -0.2$. Hence, the solution or the intersection is $(1.2, -0.2)$.

Check:
For the first equation: $8(\mathbf{1.2}) + 3(\mathbf{-0.2}) = 9.6 - 0.6 = 9$ is true.
For the second equation: $\mathbf{-0.2} = 4(\mathbf{1.2}) - 5 = 4.8 - 5 = -0.2$ is true.

Example 2: Solve the system of equations by substitution:
$$y = 2x - 6$$
$$y = 2.1x - 11.2$$

Solution: In this problem, both equations have the variable y by itself. Since y **equals** $2x - 6$ in the first equation, we can take that and replace y in the second equation with $2x - 6$. So, $\mathbf{y = 2.1x - 11.2}$ becomes $\mathbf{2x - 6} = 2.1x - 11.2$. Adding 6 to both sides, we have

$2x - 6 + 6 = 2.1x - 11.2 + 6$
=> $\quad 2x = 2.1x - 5.2 \quad$ Simplify
=> $\quad 2x - 2.1x = 2.1x - 2.1x - 5.2 \quad$ Subtract 2.1x to both sides
=> $\quad -0.1x = -5.2 \quad$ Simplify
=> $\quad x = \dfrac{-5.2}{-0.1} = 52 \quad$ Divide by –0.1

After getting $x = 52$, we can plug it back into either one of the orignal equations to solve for y. But we should use the first equation to avoid decimals. Thus, we have $y = 2(52) - 6 = 104 - 6 = 98$. Hence, the solution or the intersection is $(52, 98)$. Although the intersection is not a decimal, it would still be tricky to identify it on the graph because of the scale.

Example 3: Solve the system of equations by substitution:
$$5x + 9y = 23$$
$$x - 7y = 15$$

Solution: In this problem, neither equations have a variable by itself. So, we have to choose one of the equations and solve for one of the variables. Since there is no number in front of x in the second equation, choosing this equation to solve for x in terms of y is the easiest way to go. So given x – 7y = 15, adding 7y to both sides produces x – 7y + 7y = 7y + 15. Simplifying gives us **x = 7y + 15**. Now we have an equation with x by itself. We can substitute x with 7y + 15 in the first equation. As a result, 5x + 9y = 23 becomes 5**(7y + 15)** + 9y = 23.

=>	35y + 75 + 9y = 23	Distribute
=>	44y + 75 = 23	Combine like terms
=>	44y + 75 – 75 = 23 – 75	Subtract 75 to both sides
=>	44y = –52	Simplify
=>	$y = \dfrac{-52}{44} = \dfrac{-13}{11}$	Divide 44 to both sides

Although we can plug $y = \dfrac{-13}{11}$ back into either one of the orignal equations to solve for x, remember it is better to use the equation with x by itself (i.e., x = 7y + 15). Thus, we have $x = 7(\dfrac{-13}{11}) + 15 = \dfrac{-91}{11} + 15 = \dfrac{-91}{11} + \dfrac{165}{11} = \dfrac{74}{11}$. Hence, the solution or the intersection is $(\dfrac{74}{11}, \dfrac{-13}{11})$.

Example 4: Solve the system of equations by substitution:
$$3x - 12y = 7$$
$$5x + 8y = 4$$

Solution: In this problem, neither equations have a variable by itself. Also, neither equations have a variable with no number in front of it. So, we have to randomly choose one of the equations and solve for one of the variables. Let's solve the first equation for x in terms of y. Given 3x – 12y = 7, we have

	3x – 12y + 12y = 12y + 7	Add 12y to both sides
=>	3x = 12y + 7	Simplify
=>	**x = 4y + $\dfrac{7}{3}$**	Divide by 3 to both sides

Now we have an equation with x by itself. We can substitute x with $4y + \dfrac{7}{3}$ in the second equation. As a result, 5x + 8y = 4 becomes

Pre-Algebra

$$5(4y + \frac{7}{3}) + 8y = 4.$$

$$\Rightarrow \quad 20y + \frac{35}{3} + 8y = 4 \qquad \text{Distribute}$$

$$\Rightarrow \quad 28y + \frac{35}{3} = 4 \qquad \text{Simplify}$$

$$\Rightarrow \quad 28y + \frac{35}{3} - \frac{35}{3} = 4 - \frac{35}{3} \qquad \text{Subtract } \frac{35}{3}$$

$$\Rightarrow \quad 28y = \frac{12}{3} - \frac{35}{3} = \frac{-23}{3} \qquad \text{Simplify}$$

$$\Rightarrow \quad y = \frac{-23}{3} \times \frac{1}{28} = \frac{-23}{84} \qquad \text{Multiply by } \frac{1}{28}$$

Now plug $y = \frac{-23}{84}$ into $x = 4y + \frac{7}{3}$, we have $x = 4(\frac{-23}{84}) + \frac{7}{3} = (\frac{-23}{21}) + \frac{7}{3} = (\frac{-23}{21}) + \frac{49}{21} = \frac{26}{21}$. Hence, the solution or the intersection is $(\frac{26}{21}, \frac{-23}{84})$.

Example 5: Solve the system of equations by substitution:
$$14x - 2y = 1$$
$$y + 5 = 7x - 3$$

Solution: Since there is no number in front of y in the second equation, solve it for y in terms of x. Subtract 5 to both sides, we have **y = 7x – 8**. Now substitute y with 7x – 8 in the first equation, **14x – 2y = 1** becomes 14x – 2**(7x – 8)** = 1. Next,
we have 14x – 14x + 16 = 1   Distribute
$\Rightarrow$ 16 = 1   Simplify

At this point, we have an equation with no variable. Now since the statement 16 = 1 is **false**, this means there is **no solution**. From the previous lesson, we also know that, graphically, this means the two lines are parallel.

Note that instead of getting a false statement, suppose we get a **true** statement such as 10 = 10, this would mean that there are **infinitely many solutions**. The two lines are actually one (i.e., they intersect infinitely many times).

Example 6: The sum of two integers is 143 and the difference is 9. Find the integers.

Solution: Let x be the first integer. Let y be the second integer. So,

$$x + y = 143, \text{ and}$$
$$x - y = 9.$$

We can solve the second equation for x in terms of y. Adding y to both sides, $x - y = 9$ becomes $x - y + y = y + 9$. Simplifying gives us **x = y + 9**. Substitute it into the first equation, we have $(y + 9) + y = 143$, or $2y + 9 = 143$. Subtract 9 to both sides produces $2y = 134$. Divide by 2 gives **y = 67**. Plug it into $x = y + 9$, we have $x = 67 + 9 = 76$. Hence, the two integers are 76 and 67.

Practice 18

Solve each system of equations by substitution.

1) $y = 3x + 4$
   $2x + y = 9$

2) $3y - 5x = -9$
   $x = -2y + 7$

3) $y = 2x - 8$
   $y = x + 5$

4) $2x + 2y = 11$
   $2x - 2y = 9$

5) $4y = 8 - 6x$
   $x + 2y = -1$

6) $y + 10 = 0.5x$
   $3x - 6y = 12$

7) $3x + 2y = 4$
   $4x - 6y = 1$

8) $y = 13.5 - 2x$
   $4x + 2y = 27$

9) $3x = -2$
   $5y = 4 + 6x$

10) $y = 13x + 3$
    $x = 13y + 45$

11) $3x + 4y = 5$
    $6x + 8y = 7$

12) $x + 5y = 0$
    $4x - 10y = -21$

13) $12 = 2y + 3x$
    $6x + 4y = 8$

14) $0.03x = -6y + 9$
    $0.01x + 5y = 12$

15) $4y = 2 + 3x$
    $2x = 2y - 3$

16) $2y - 8x = 2x - 8y + 1$
    $20y - 19x = 0$

17) $y = 4(2x - 3.1)$
    $y = 2 - 8x$

18) $y = \frac{2}{9}x + 6$
    $27y + 6 = 6x$

19) Twice the first number plus the second number is 40. Five times the first number minus the second number is 30. Find the numbers.

20) The sum of two numbers is −50 and the difference is −5. Find the numbers.

Lesson 19: Laws of Exponents

In lesson 2, you learned that we can use an exponent as a short cut for writing repeated multiplications. For example, $2 \times 2 \times 2 \times 2 \times 2$ can be written as 2^5 and vice versa. This means that if we have $2^3 \times 2^4$, we can think of it as $(2 \times 2 \times 2) \times (2 \times 2 \times 2 \times 2) = 2^7$.

In general, here is the law for the **product of powers**: $a^m \times a^n = a^{m+n}$.

Example 1: Simplify. Express using exponents.

    a) $5^2 \cdot 5^4$ b) $x^3 \cdot x^2 \cdot x \cdot x^5$

    c) $3y^2 \cdot 7y^5$ d) $8mn^3 \cdot 6m^2 \cdot 2m^3n^4$

Solution: a) This problem is quite simple if you know the rule clearly. $5^2 \cdot 5^4 = 5^6$. You simply add the exponents. However, if you are over thinking or second guessing yourself, you can easily make a mistake. Consider this as a multiple choice question: $5^2 \cdot 5^4 =$ ___.
         a. 25^8 b. 25^6 c. 5^8 d. 5^6
If you are not careful, you may think all of these choices look "good". But, only (d) is correct.

b) $x^3 \cdot x^2 \cdot x \cdot x^5 = x^{3+2+1+5} = x^{11}$. Note that the exponent for x is 1 not 0, because $x = x^1$.

c) $3y^2 \cdot 7y^5 = 3 \cdot 7 \cdot y^2 \cdot y^5 = 21y^{2+5} = 21y^7$. Note that you multiply the numbers in the front, but you add the exponents.

d) $8mn^3 \cdot 6m^2 \cdot 2m^3n^4 = 8 \cdot 6 \cdot 2 \cdot m^{1+2+3} \cdot n^{3+4} = 96m^6n^7$.

The next question is, what about dividing powers? For example, what is $2^8 \div 2^2$? Using the meaning of exponents, we can write $\dfrac{2^8}{2^2}$ as $\dfrac{2 \cdot 2 \cdot 2 \cdot 2 \cdot 2 \cdot 2 \cdot 2 \cdot 2}{2 \cdot 2}$. Simplifying the fraction gives $2 \cdot 2 \cdot 2 \cdot 2 \cdot 2 \cdot 2 = 2^6$.

So in general, the law for the **quotient of powers** is: $\dfrac{a^m}{a^n} = a^{m-n}$.

Example 2: Simplify. Express using exponents.

a) $\dfrac{9^{18}}{9^6}$

b) $\dfrac{a^6 \cdot a^8}{a^5 \cdot a^2}$

c) $\dfrac{6x^9y^2}{14x^3yz}$

d) $\dfrac{p^7}{p^7}$

Solution: a) $\dfrac{9^{18}}{9^6} = 9^{18-6} = 9^{12}$.

b) First do the numerator and denominator separately using the law for the product of powers. Then use the law for the quotient of powers to simplify the results. Hence,

$$\dfrac{a^6 \cdot a^8}{a^5 \cdot a^2} = \dfrac{a^{6+8}}{a^{5+2}} = \dfrac{a^{14}}{a^7} = a^{14-7} = a^7.$$

c) First reduce the fraction $\dfrac{6}{14}$. Then use the law for the quotient of powers to simplify the variables x and y. Leave the variable z alone since it only appears in the denominator. Hence,

$$\dfrac{6x^9y^2}{14x^3yz} = \dfrac{6}{14} \cdot \dfrac{x^9}{x^3} \cdot \dfrac{y^2}{y} \cdot \dfrac{1}{z} = \dfrac{3}{7} \cdot x^6 \cdot y^1 \cdot \dfrac{1}{z} = \dfrac{3x^6y}{7z}.$$

d) Using the law for the quotient of powers, $\dfrac{p^7}{p^7} = p^{7-7} = p^0$.

But, $\dfrac{p^7}{p^7}$ also equals 1 since the numerator and denominator are the same. Hence, $\dfrac{p^7}{p^7} = p^0 = 1$.

From the result in example 2d, we see that the law for **zero exponent** is $a^0 = 1$.

What about negative exponent?

In fact, we can use the law for the quotient of powers to derive the law for negative exponent as well. For example, $\dfrac{x^4}{x^7} = x^{4-7} = x^{-3}$.

But $\dfrac{x^4}{x^7}$ also equals $\dfrac{x \cdot x \cdot x \cdot x}{x \cdot x \cdot x \cdot x \cdot x \cdot x \cdot x} = \dfrac{1}{x \cdot x \cdot x} = \dfrac{1}{x^3}$. Hence, $x^{-3} = \dfrac{1}{x^3}$.

In general, the law for **negative exponent** is $a^{-m} = \dfrac{1}{a^m}$.

Also, $\dfrac{1}{a^{-m}} = \dfrac{1}{\frac{1}{a^m}} = 1 \div \dfrac{1}{a^m} = 1 \times \dfrac{a^m}{1} = a^m$.

In summary, if the negative exponent is in the numerator such as a^{-m}, you move it to the denominator in order to make it positive. On the other hand, if the negative exponent is already in the denominator such as $\dfrac{1}{a^{-m}}$, you move it to the numerator if you want to make it positive.

Example 3: Simplify. Express answers using only positive exponents.

    a) $\dfrac{2^3}{2^{10}}$ b) $\dfrac{m^{-4}n^8}{m^5 n^2}$ c) $\dfrac{-24x^2 y^{-3}}{16x^{-2} yz^{-3}}$

Solution: a) $\dfrac{2^3}{2^{10}} = 2^{3-10} = 2^{-7} = \dfrac{1}{2^7} = \dfrac{1}{128}$.

     b) $\dfrac{m^{-4}n^8}{m^5 n^2} = m^{-4-5} \cdot n^{8-2} = m^{-9} \cdot n^6 = \dfrac{n^6}{m^9}$.

Alternatively, you can also do the following:
$\dfrac{m^{-4}n^8}{m^5 n^2} = \dfrac{n^8}{m^4 m^5 n^2} = \dfrac{n^{8-2}}{m^{4+5}} = \dfrac{n^6}{m^9}$.

c) After reducing $\dfrac{-24}{16}$, move all the negative exponents in the opposite direction. Then simplify using the law for the product of powers.

Pre-Algebra

Hence, $\dfrac{-24x^2y^{-3}}{16x^{-2}yz^{-3}} = \dfrac{-3x^2x^2z^3}{2yy^3} = \dfrac{-3x^4z^3}{2y^4}$.

We will conclude this lesson with one more law. It is the power of a power.

For example, what is $(2^3)^4$?

The outter exponent means $(2^3)^4 = (2^3)(2^3)(2^3)(2^3)$.

The inner exponent means $(2^3)^4 = (2^3)(2^3)(2^3)(2^3) = (2\cdot2\cdot2)(2\cdot2\cdot2)(2\cdot2\cdot2)(2\cdot2\cdot2)$.

Hence, $(2^3)^4 = (2^3)(2^3)(2^3)(2^3) = (2\cdot2\cdot2)(2\cdot2\cdot2)(2\cdot2\cdot2)(2\cdot2\cdot2) = 2^{12}$.

In general, the law for the **power of a power** is $(a^m)^n = a^{m\times n}$.

Example 4: Simplify. Express answers using only positive exponents.

$$\text{a) } (3^4)^{-5} \qquad \text{b) } \dfrac{(a^2)^6(5b^3)^2}{(a^5)^4(b^2)^3} \qquad \text{c) } \dfrac{-(x^{-4})^0(4y^9)^2z}{(x^{-1})^4(y^{-5})^3z^{-1}}$$

Solution: a) $(3^4)^{-5} = 3^{4\times-5} = 3^{-20} = \dfrac{1}{3^{20}}$.

b) $\dfrac{(a^2)^6(5b^3)^2}{(a^5)^4(b^2)^3} = \dfrac{a^{12}25b^6}{a^{20}b^6} = 25a^{12-20} = 25a^{-8} = \dfrac{25}{a^8}$.

Note that it is a common mistake for students to say $(5b^3)^2 = 5b^6$. $(5b^3)^2 = (5b^3)(5b^3) = 25b^6$.

c) $\dfrac{-(x^{-4})^0(4y^9)^2z}{(x^{-1})^4(y^{-5})^3z^{-1}} = \dfrac{-x^0 16y^{18}z}{x^{-4}y^{-15}z^{-1}} =$

$-16x^{0--4}y^{18--15}z^{1--1} = -16x^4y^{33}z^2$.

Practice 19

Simplify. Express answers using only positive exponents.

1) $x^6 \cdot x^2$

2) $(x^6)^2$

3) $(-3y^{-5})(-y)(2y^3)^2$

4) $\dfrac{p^9 q^{15}}{p^8 q^{10}}$

5) $\dfrac{6a^{-1}b^{-2}}{30a^{-5}b^2 c}$

6) $\dfrac{x^0 z^{-6}}{x^{-3} y^{-7}}$

7) $(2a)^4 \cdot a^{-2} \cdot a^4 \cdot \dfrac{1}{a^6}$

8) $\dfrac{(-2x^3)^5 (y^{-8})^0}{(x^2)^6 (y^3)^3 z^{-1}}$

9) $\dfrac{x^{20} y z^{-30}}{x^{17} y^{-9} z^{-25}}$

10) $[(x^3)^5 \cdot (x^2)]^0$

11) $y^{-1} \cdot y^{-2} \cdot \dfrac{y^3}{y^2}$

12) $(x^5 \cdot x^{-8} \cdot x)^{-1}$

13) $\left(\dfrac{1}{x^{-3} x^{-7}}\right)^{-2}$

14) $\dfrac{2^{-2}}{3^{-3}}$

15) $\dfrac{(10^2)^3}{2^{-3} \times 10^4}$

16) $\dfrac{\frac{x^2 + y^3}{xy^3}}{\frac{2}{x^3 y^5}}$

17) $\dfrac{(-x)^{-5} y}{(-x)^{-3}(-y)^{-4}}$

18) $\left(\dfrac{1}{2^{-1}} - \dfrac{2}{2^{-2}} - \dfrac{3}{2^{-3}}\right)^2$

19) $\dfrac{a-b}{a^{-5} b^{-7}}$

20) $(2x^3)^2 \left[2x - 4 - \dfrac{1}{(2x)^3}\right]$

Bonus: Simplify $\left(\dfrac{1}{x} - \dfrac{1}{y} - \dfrac{1}{z}\right)^{-1}$

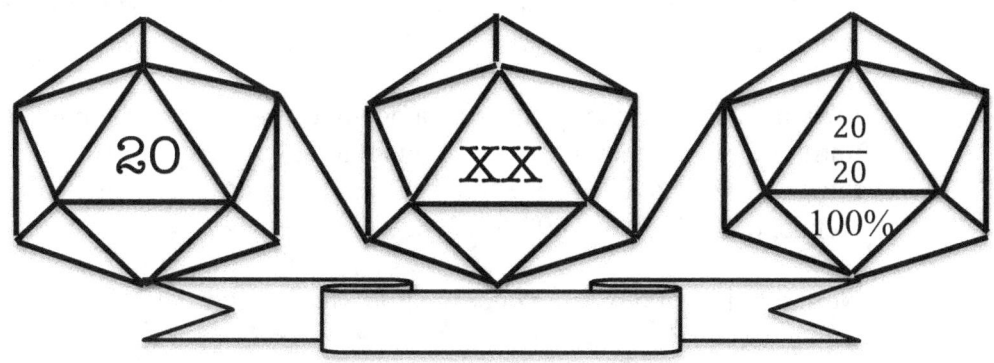

Lesson 20: Scientific Notation

Scientists sometimes have to work with extremely large or extremely small numbers. For example, the mass of the earth is approximately 5,972,000,000,000,000,000,000,000 kg. The radius of a hydrogen atom is approximately 0.000000000037 m. In **standard form**, not only these numbers are hard to read and write, they are hard to use in computations. As a results, scientists use **scientific notation** to deal with these numbers.

In scientific notation, a number can be written in the form of $a \times 10^c$, $1 \leq a < 10$ and c is an integer.

For example, the mass of the earth in scientific notation is 5.972×10^{24} kg, and the radius of a hydrogen atom is 3.7×10^{-11} m.

Example 1: Write each number in scientific notation.

    a) 243,700,000   b) 0.00000598

    c) 145   d) 379×10^{14}

Solution: a) First place a decimal at the end of the number and move it to the **left** until you get a number greater or equal to 1, but less than 10. In this case, you have to move the decimal 8 times to the left. Hence, $243,700,000 = 2.437 \times 10^8$.

Note that you do not have to write the trailing (or leading) zeros after you move the decimal. That is why the answer is NOT being written as 2.43700000×10^8.

b) This number already contains a decimal. Move it to the **right** until you get a number greater or equal to 1, but less than 10. In this case, you have to move the decimal 6 times to the right. Hence, $0.00000598 = 5.98 \times 10^{-6}$.

Note that if you move the decimal point to the right, the exponent is negative and vice versa.

c) 145 is the same as 145.0, so $145 = 1.45 \times 10^2$.

d) Even though 379×10^{14} looks like a number written in scientific notation, it is NOT, since 379 is greater than 10. We have to turn 379 into 3.79×10^2.
Hence, $379 \times 10^{14} = (3.79 \times 10^2) \times 10^{14} = 3.79 \times 10^{16}$.

Example 2: Write each number in standard notation.

a) 2.033×10^4 b) 8.76×10^{-5}

c) 1×10^{-1} d) 3.1415×10^3

Solution: a) For these problems, simply move the decimal to the right if the exponent is positive and to the left if it is negative. Here since we have to move the decimal 4 times to the right, we must add an extra zero to the answer. Hence, $2.033 \times 10^4 = 20330$.

b) Move the decimal 5 times to the left and add leading zeros. Hence, $8.76 \times 10^{-5} = 0.0000876$.

c) 1 is the same as 1.0. Move the decimal 1 times to the left. Hence, $1 \times 10^{-1} = 0.1$.

d) Move the decimal 3 times to the right. Hence, $3.1415 \times 10^3 = 3141.5$.

We can use the laws of exponents from the previous lesson to multiply and divide numbers that are written in scientific notation.

Pre-Algebra

Example 3: Multiply or divide. Express the answer in scientific notation.

a) $(2.3 \times 10^7) \cdot (5.0 \times 10^{-2})$ b) $\dfrac{9.6 \times 10^{11}}{3.2 \times 10^6}$

c) $(1.2 \times 10^{-3}) \cdot (4.1 \times 10^{-4})$ d) $\dfrac{(2.7 \times 10^{-1})(8 \times 10^{-2})}{5.4 \times 10^{-7}}$

Solution: a) First multiply 2.3 and 5.0 to get 11.5. Then use the law for the product of powers to multiply 10^7 and 10^{-2} to get 10^5. Finally, write 11.5×10^5 in scientific notation to get 1.15×10^6. In other words, $(2.3 \times 10^7) \cdot (5.0 \times 10^{-2}) = (2.3 \times 5.0) \cdot (10^7 \times 10^{-2})$ $= 11.5 \times 10^{7-2} = 11.5 \times 10^5 = 1.15 \times 10^6$.

b) First divide 9.6 by 3.2 to get 3. Then use the law for the quotient of powers to divide 10^{11} by 10^6 to get 10^5. Hence, the answer is 3×10^5, since it is already in scientific notation.

c) $(1.2 \times 10^{-3}) \cdot (4.1 \times 10^{-4}) = (1.2 \times 4.1) \cdot (10^{-3} \times 10^{-4})$ $= 4.92 \times 10^{-3-4} = 4.92 \times 10^{-7}$.

d) $\dfrac{(2.7 \times 10^{-1})(8 \times 10^{-2})}{5.4 \times 10^{-7}} = \left(\dfrac{2.7 \times 8}{5.4}\right) \cdot \left(\dfrac{10^{-1} \times 10^{-2}}{10^{-7}}\right) =$

$4 \times 10^{-1 + -2 - -7} = 4 \times 10^4$. Note that you do not need to use a calculator because $\dfrac{2.7}{5.4}$ is $\dfrac{1}{2}$, and $\dfrac{1}{2} \times 8 = 4$.

Example 4: Order the following numbers from least to greatest.

8.4×10^6, 967540, 0.599×10^7, 0.00123×10^{10}.

Solution: The first number is already in scientific notation. Write the other three numbers in the same form in order to compare them.
Here they are: $967540 = \mathbf{9.6754 \times 10^5}$, $0.599 \times 10^7 = \mathbf{5.99 \times 10^6}$, $0.00123 \times 10^{10} = \mathbf{1.23 \times 10^7}$.

Since $\mathbf{9.6754 \times 10^5} < \mathbf{5.99 \times 10^6} < 8.4 \times 10^6 < \mathbf{1.23 \times 10^7}$, we have $967540 < 0.599 \times 10^7 < 8.4 \times 10^6 < 0.00123 \times 10^{10}$.

20/20 Math

Practice 20

Write each number in scientific notation.

1) 39400000

2) 0.000234

3) 99

4) 102×10^{-12}

5) 0.012×10^9

6) $\dfrac{1}{2000}$

Write each number in standard notation.

7) 1.23×10^7

8) 5.003×10^{-6}

9) 99.8×10^{-4}

10) 0.00046×10^{10}

Multiply or Divide. Express the answers in scientific notation.

11) $(4.1 \times 10^9) \cdot (2.2 \times 10^{-8})$

12) $\dfrac{16.4 \times 10^{18}}{4.1 \times 10^{12}}$

13) $\dfrac{(6 \times 10^{12})(8 \times 10^{-2})}{96 \times 10^5}$

14) $\dfrac{(1.25 \times 10^4)(6.4 \times 10^{-8})}{(8 \times 10^{-10})(0.25 \times 10^{-2})}$

15) $\dfrac{(9 \times 10^7)}{(2 \times 10^8)(1.5 \times 10^6)}$

16) $(6.5 \times 10^3)^2$

17) $\dfrac{(6 \times 10^4)^2}{4 \times 10^2}$

18) $\dfrac{(2 \times 10^{-17})}{(5 \times 10^{11})^{-2}}$

19) $(2 \times 10^{-2}) \cdot (5.5 \times 10^4)^2$

Order the numbers from least to greatest.

20) $3532200,\ 21.6 \times 10^5,\ 0.9 \times 10^7,\ 135 \times 10^4$.

Answers for Even Problems

Practice 1

2a. IR, RE	2b. IR, RE	2c. R, RE
4. 792	6. 562.5	8. 0
10. 72.25	12. 5	14. 3690
16. 570	18. 11025	20. 41.05

Practice 2

2. 180   4. 12.5   6. $\dfrac{60}{49}$ 8. 64

10. <   12. =   14. >   16. $[2 + (3 + 4) \times 5] - 6$

18. $[(16 + 8) \div 6] \times 4$ 20. $3 + (3 + 3) \times 3 - 3 \div 3$

Practice 3

2. −19   4. 90   6. −41   8. 14   10. 72.9

12. 6.25   14. 43   16. −100   18. <   20. <

Practice 4

2. yes   4. no   6. no   8. no   10. no

12. 91 is not divisible by 2, 3, 4, 5, 6, 8, 9, or 10

14. 964 is divisible by 2, 4

16. 10000 is divisible by 2, 4, 5, 8, 10

18. 52800   20. 540

Practice 5

2. $2^6 \times 3 \times 5$ **4.** $2 \times 3^2 \times 7 \times 11$ **6.** 56 **8.** 291

10. 320 **12.** 546 **14.** 14

16. 13 **18.** 9 **20.** 27

Practice 6

2a. 6.5454… **2b.** 3.625 **2c.** 3.66666…

4. 1.8 pound per dollar **6.** 85% **8.** $41\frac{1}{3}\%$

10. 70% **12.** 0.67 **14.** 215%

16. 70% **18.** 72 **20.** 102

Practice 7

2. $\frac{9}{8}$ **4.** 52 **6.** 666 **8.** −48 **10.** 90.25

12. 100 **14.** 0.0056 **16.** 360 **18.** 63 **20.** 888.25

Practice 8

2. 0.0010023 kg **4.** 10 km **6.** 35000000 **8.** 222.22…

10. 0.0382 g **12.** 0.64 g, 80 cg, 7400 mg, 0.0075 kg

14. 99.8 m **16.** 30 km

18. 65 kiloliter per year **20.** 0.17 cents per millileter

Pre-Algebra

Practice 9

2. 144 in. **4.** 9 ft **6.** $13\frac{8}{9}$ yd **8.** 6 oz **10.** 3168 oz

12. 163.2 c **14.** 4.5 gal **16.** 384 fl oz **18.** 4 gal **20.** 35.2 pt

Practice 10

2. 257 **4.** 680 **6.** 121.9 **8.** $\frac{5}{18}$

10. $-a^2b - 3ab^2 - 2ab - 3a - 3b$ **12.** $-1 + \frac{2}{3}xyz$ **14.** $-2x^2 + 300x - 114$

16. $-5m^2 - 4m - 5$ **18.** 441 **20.** 222.6

Practice 11

2. -3 **4.** $\frac{4}{7}$ **6.** 78 **8.** 3 **10.** $-14\frac{5}{6}$

12. 0 **14.** 225 **16.** $\frac{4}{3}$ **18.** 54 **20.** 39

Practice 12

2. no solution **4.** infinitely many solutions **6.** 0

8. infinitely many solutions **10.** 1 **12.** $x = \frac{y-b}{m}$

14. $m = \frac{E}{c^2}$ **16.** $z = \frac{4xy-1}{3xy}$

18. $w = \frac{V}{lh}$ **20.** $y = \frac{c-ax}{b}$

Practice 13

2. 28 **4.** 18 **6.** 70 **8.** 38 **10.** 62.5 hours

12. 68 cm **14.** 204 **16.** $150 **18.** 49 min **20.** 28 m

Practice 14

2. x ≤ −2
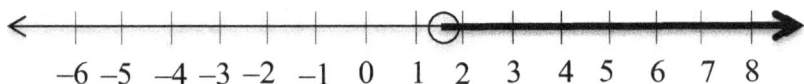

4. x > 1.5
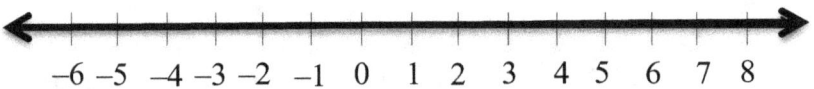

6. infinitely many solutions
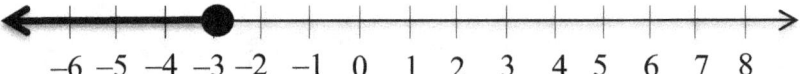

8. x ≤ −3

10. x ≤ 0
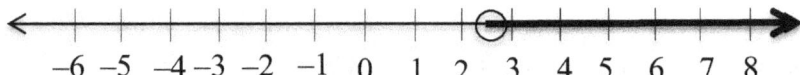

12. x > 2.5

14. x ≤ 0.5
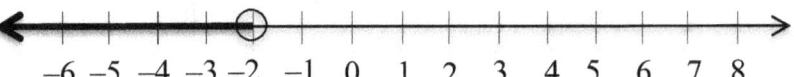

16. x < −2

18. no solution   **20.** 1, 3, 5; 3, 5, 7; 5, 7, 9

Pre-Algebra

Practice 15

2.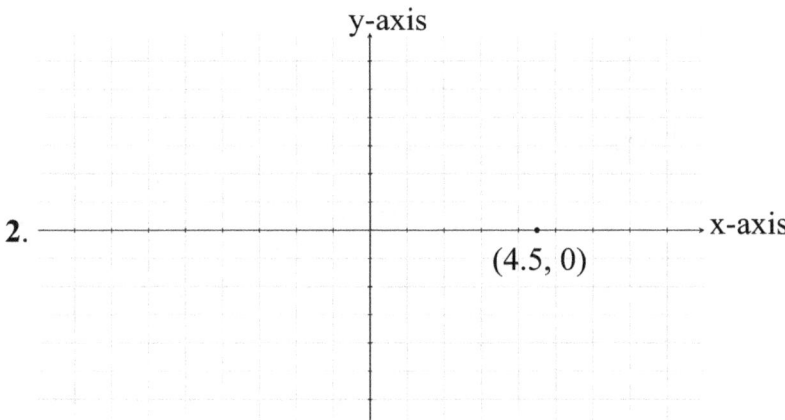

4.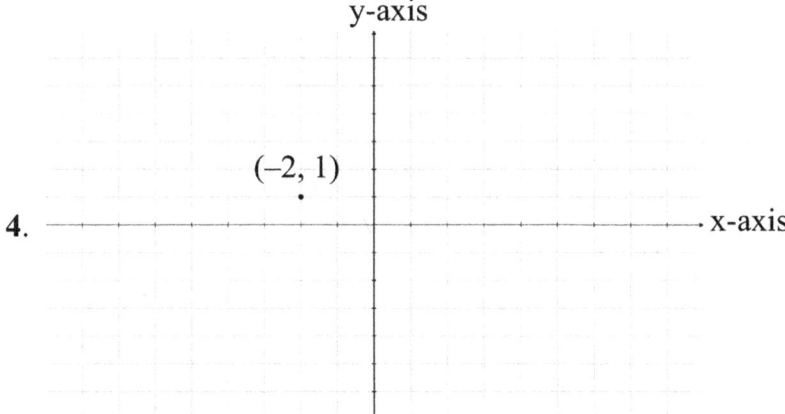

6. on the x-axis **8.** quadrant II **10.** quadrant I

12.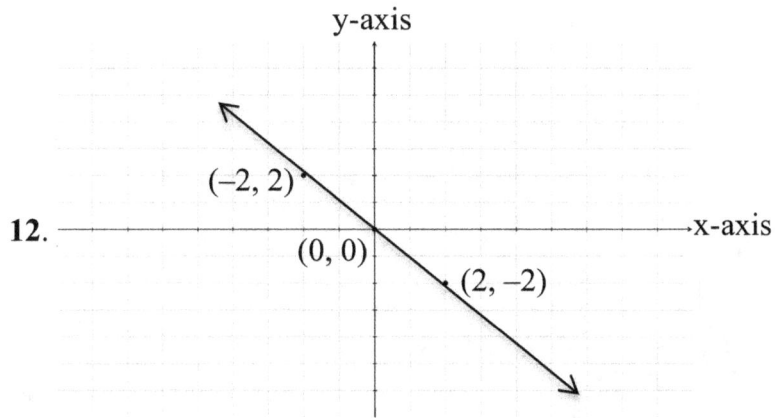

14.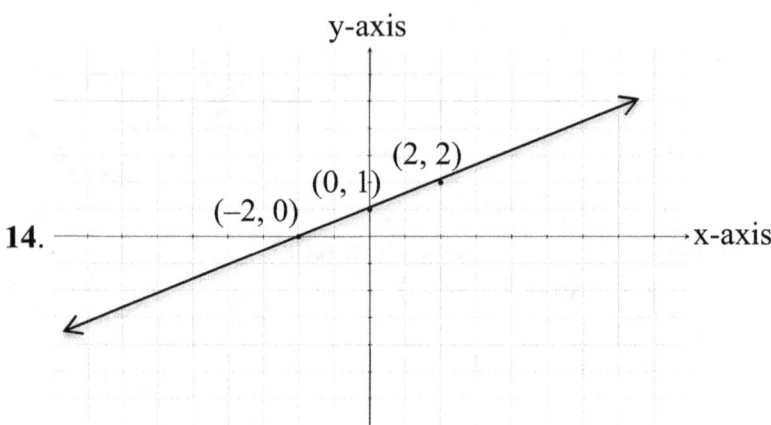

16.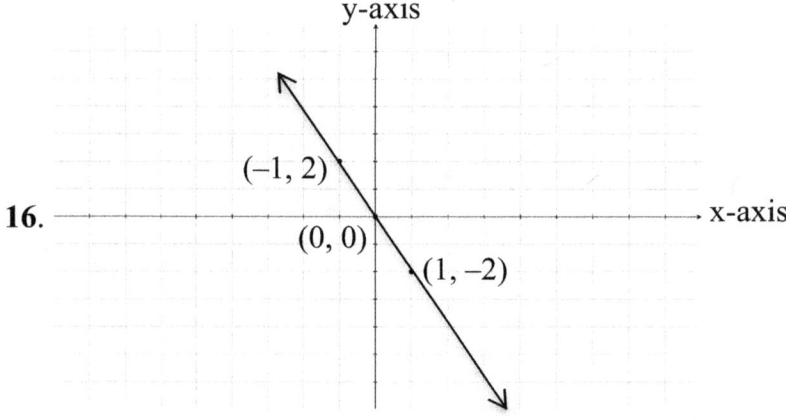

18.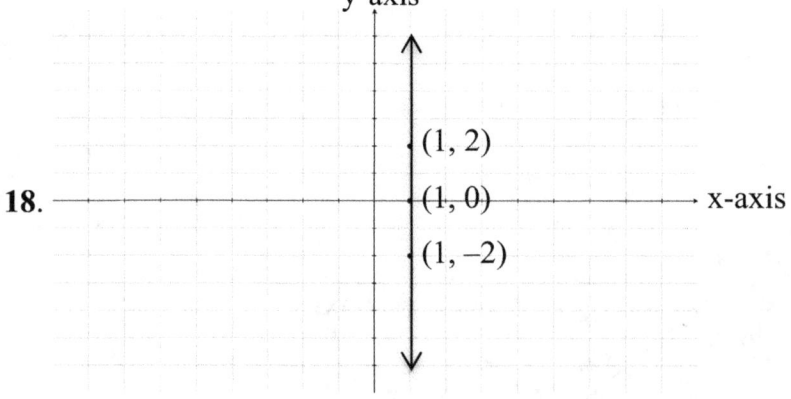

20.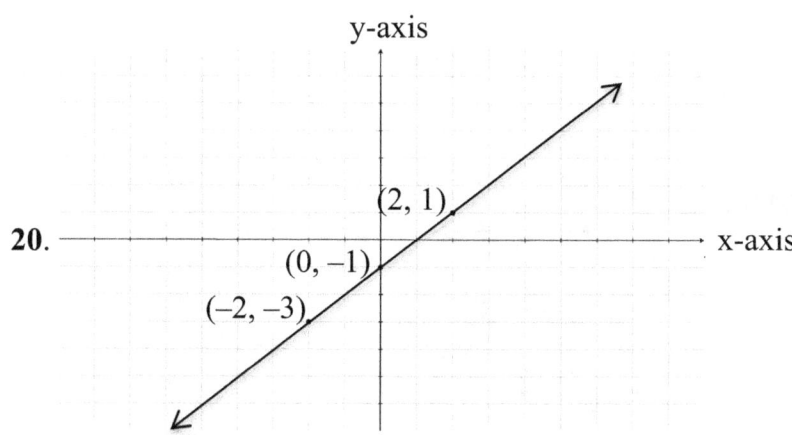

Practice 16

2.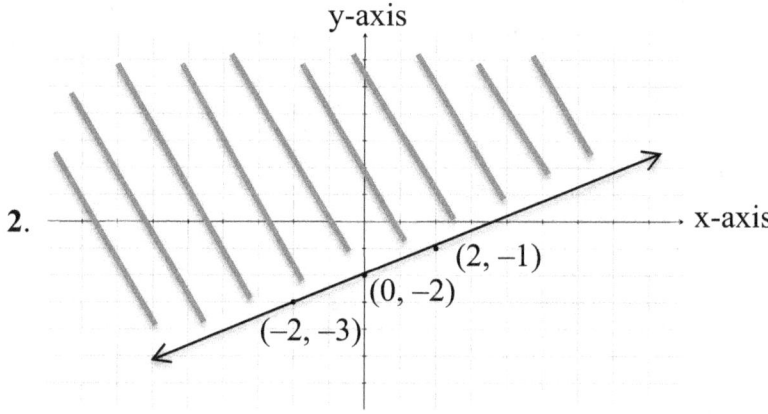

4.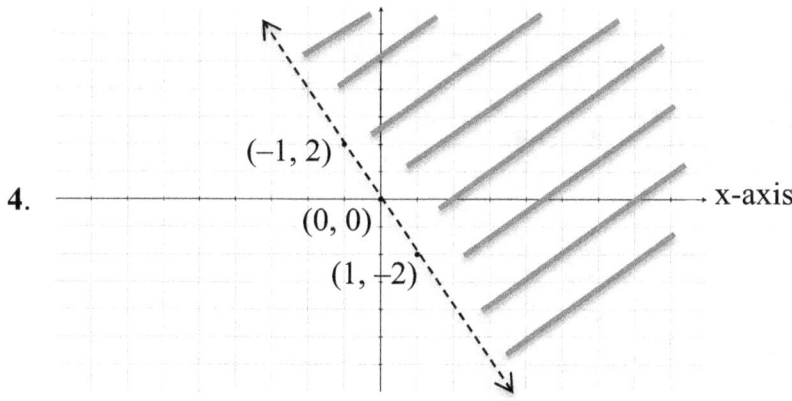

6.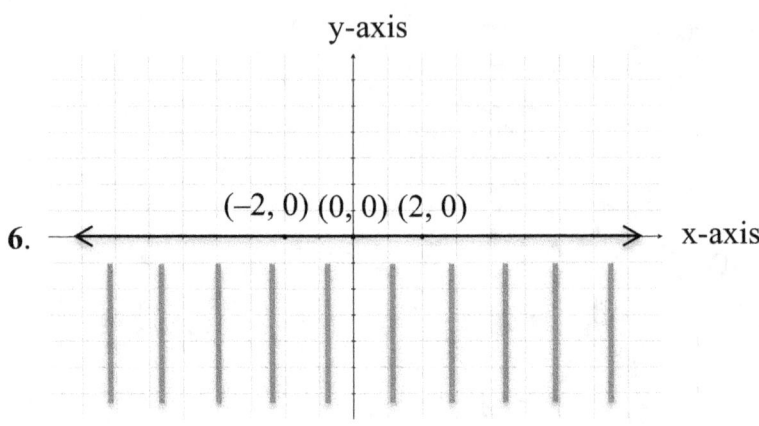

8.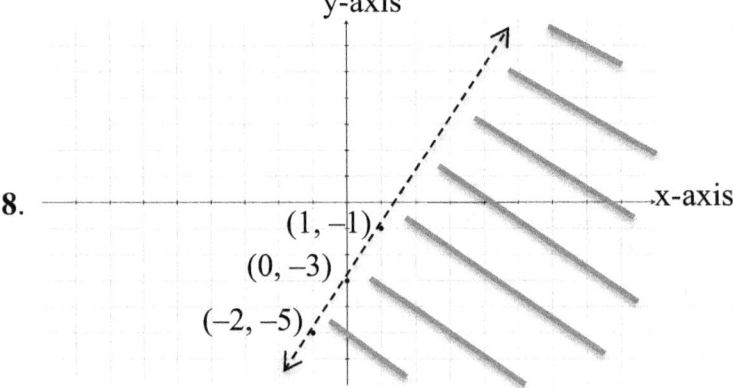

10.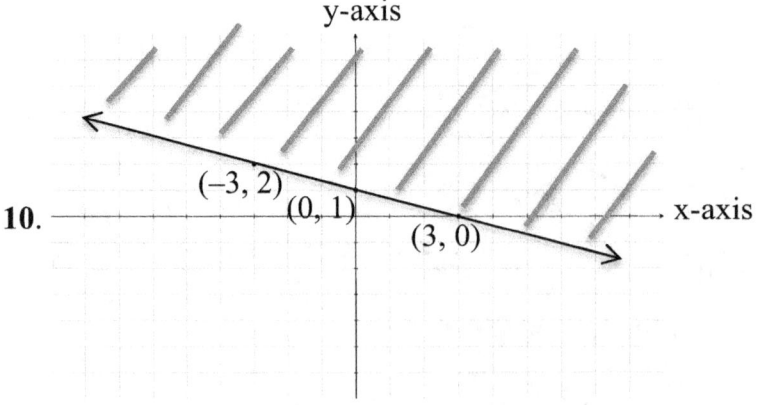

12.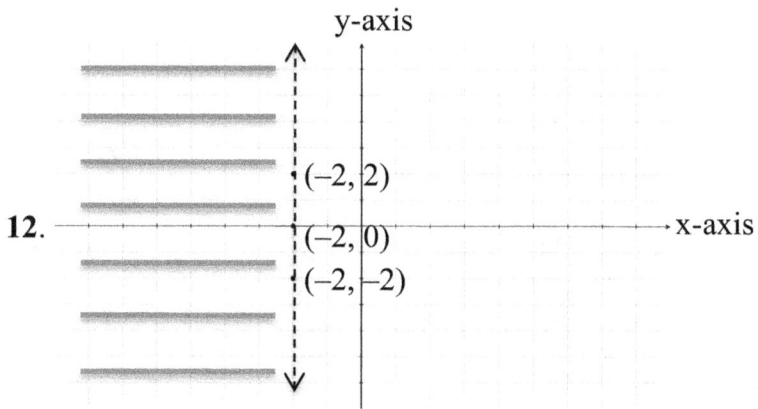

14.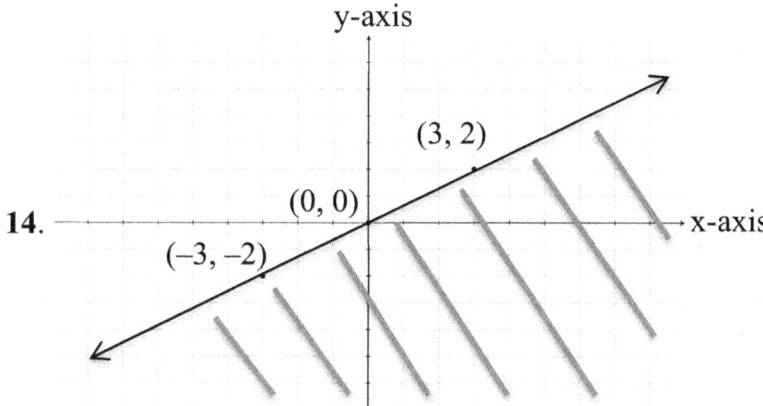

16.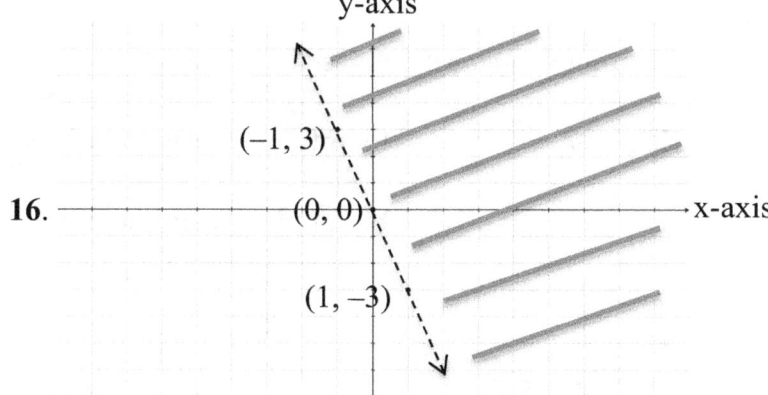

18.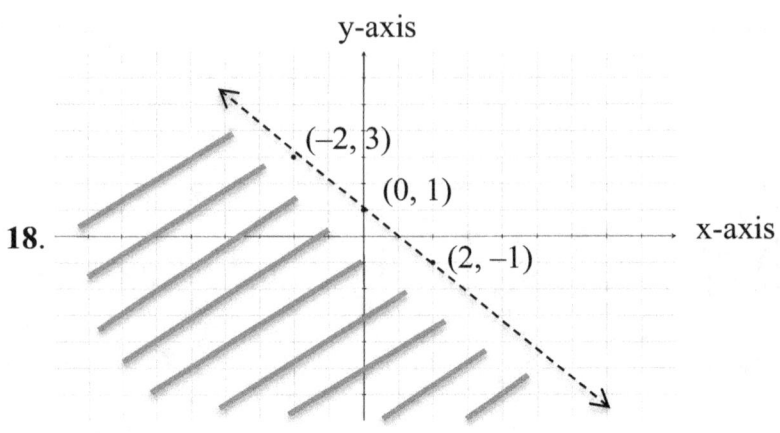

20.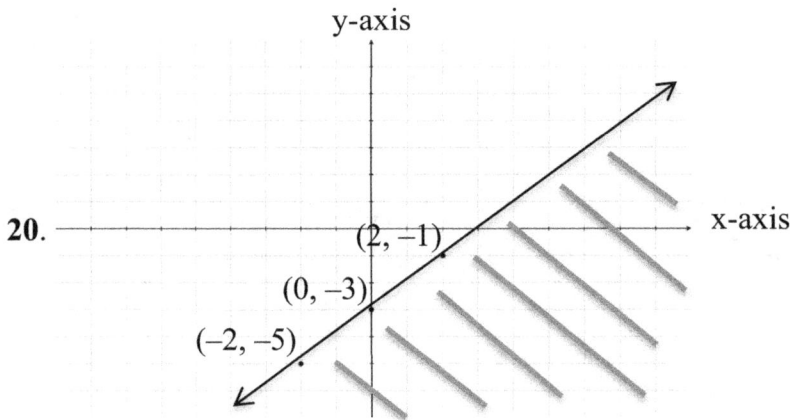

Practice 17

2. (2, 0) 4. (–3, 4) 6. infinitely many solutions 8. (0, 2)

10. no solution 12. (3, –2) 14. infinitely many solutions

16. (0, 0) 18. (5, 3) 20. infinitely many solutions

Practice 18

2. (3, 2) 4. (5, 0.5) 6. no solution

8. infinitely many solutions 10. (–0.5, –3.5) 12. (–3.5, 0.7)

14. (–300, 3)   **16.** (–2, –1.9)   **18.** no solution   **20.** (–27.5, –22.5)

Practice 19

2. x^{12} **4.** pq^5 **6.** $\dfrac{x^3 y^7}{z^6}$ **8.** $\dfrac{-32x^3 z}{y^9}$

10. 1   **12.** x^2 **14.** $\dfrac{27}{4}$ **16.** $\dfrac{x^4 y^2 + x^2 y^5}{2}$

18. 900   **20.** $8x^7 - 16x^6 - \dfrac{x^3}{2}$

Practice 20

2. 2.34×10^{-4} **4.** 1.02×10^{-10} **6.** 5×10^{-4}

8. 0.000005003   **10.** 4600000   **12.** 4×10^6

14. 4×10^8 **16.** 4.225×10^7 **18.** 5×10^6

20. 135×10^4, 21.6×10^5, 3532200, 0.9×10^7

www.ingramcontent.com/pod-product-compliance
Lightning Source LLC
Chambersburg PA
CBHW080459220526
45465CB00006B/2323